Cultural Tourism

Cultural Tourism

Editor

Kornelia Marshall

Cultural Tourism

Edited by **Kornelia Marshall**

ISBN: 978-1-68117-248-4
Library of Congress Control Number: 2016934791

Preface

Cultural tourism (or culture tourism) is the subset of tourism concerned with a country or region's culture, specifically the lifestyle of the people in those geographical areas, the history of those people, their art, architecture, religion(s), and other elements that helped shape their way of life. Cultural tourism includes tourism in urban areas, particularly historic or large cities and their cultural facilities such as museums and theatres. It can also include tourism in rural areas showcasing the traditions of indigenous cultural communities (i.e. festivals, rituals), and their values and lifestyle, as well as niches like industrial tourism and creative tourism. It is generally agreed that cultural tourists spend substantially more than standard tourists do. This form of tourism is also becoming generally more popular throughout the world. An important feature of cultural tourism is the assumption that visiting cultural and historical sites and events, related to cultural heritage is not necessarily the main motive for the trip. In this context cultural-historical tourism is rarely implemented in a clean look and most often is combined with other traditional and specialized types of tourism. This substantial feature reveals opportunities to improve the effectiveness of national and regional tourism through the development of cultural-historical tourism - through absorption and integration of cultural-historical resources in the regional tourism product and development on this basis of a regional tourism brand.Cultural tourism is an instrument for economic development that achieves economic growth by attracting visitors outside the community-host who are motivated generally or partially by an interest in the historical, artistic, scientific or related to lifestyle and traditions reality and facts of a community, region, group or institution. Such a travel is focused on the feeling of the cultural environment, including landscapes, visual and performing arts, lifestyles, values, traditions and events.The aim of this book, Cultural Tourism, is to systematically present the role and positions of cultural tourism, as one of modern tourism industry's most dynamically developing branch, in today's global tourism market both from the theoretical and the practical point of view.

Table of Contents

Chapter 1   Sustainable Tourism – A Model Approach   1

Chapter 2   Harmonious Tourism Environment and Tourists
Perception: An Empirical Study of Mountain-Type
World Cultural Heritage Sites in China   23

Chapter 3   The Role and Importance of Cultural Tourism in
Modern Tourism Industry 35

Chapter 4   Cultural Districts, Tourism and Sustainability   69

Chapter 5   Preservation and Conservation of Rural Buildings as a
Subject of Cultural Tourism: A Review Concerning the
Application of New Technologies and
Methodologies   99

Chapter 6   Cultural Values and Sustainable Tourism Governance in
Bhutan   127

Chapter 7   Local Residents' Attitude toward Sustainable Rural
Tourism Development   149

Chapter 8   An Exploration on the Effective Factors of Tourism
Industry on Protection of the Environment in the
Historical City Ghoumas   169

Chapter 9   Study on the Authenticity of Intangible Cultural
Heritage —Take Guangdong Nanhai Boluodan Temple
Fair as an Example   181

Chapter 10   Towards Sustainable Tourism Development in Zambia:
Advancing Tourism Planning and Natural Resource
Management in Livingstone (Mosi-oa-Tunya)
Area   195

Chapter 11   Integrating Cultural and Nostalgia Tourism to Initiate A Quality Tourism Experiences at Chiangkan, Leuy Province, Thailand225

Chapter 12   Rural Tourism in the South of Spain: An Opportunity for Rural Development   235

Chapter 13   Perspectives on Cultural and Sustainable Rural Tourism in a Smart Region: The Case Study of Marmilla in Sardinia (Italy)   253

Chapter 14   The Cultural Tourism Management under Context of World Heritage Sites: Stakeholders' Opinions between Luang Prabang Communities, Laos and Muang-kao Communities, Sukhothai,Thailand ☆   277

Chapter 15   Cultural Tourism Potential, as Part of Rural Tourism Development in the North-East of Romania   291

Index   303

CHAPTER 1

Sustainable Tourism – A Model Approach

Leszek Butowski

Vistula University in Warsaw Poland

1. INTRODUCTION

Sustainable tourism, sustainable development through tourism, principles of sustainable development in tourism and tourism development in terms of sustainable tourism, in the literature often treated as names for the same phenomenon, are becoming increasingly interesting for scholars and practicians of tourism from various countries. It results from the fact that sustainable tourism (at least declaratively – on the institutional level) is considered as the most desirable form of tourism development on particular reception areas, especially those which preserved the most natural and cultural authenticity values.

Simultaneously, it should be noted that the rich scientific literature concerning sustainable tourism focuses attention primarily on descriptive presentation of its various aspects, with particular emphasis on the idea, the origins and the evolution of the phenomenon as well as terminological issues related to it. The authors pay much attention to reveal the relationships between sustainable tourism (as a form of tourism development) and particular types of tourism (as forms of tourist movement). At the same time, it should be marked that there are skeptical voices, which refer especially to the role that sustainable tourism is ascribed – as a remedy for all the problems of contemporary tourism. It also seems that, taking into consideration the hitherto scientific output related to sustainable tourism, the works devoted the theoretical aspects of sustainable tourism are in minority.

Relatively weak theoretical grounds together with the ambiguity and diversity of views on sustainable tourism as well as the descriptive approach, which dominates in the literature, have prompted the author of this article to make an attempt to render the essence of sustainable tourism in a model approach. When creating the theoretical model of sustainable tourism, the author tried to take into consideration all its main features (and interrelations ocurring among them) and to simultaneously follow certain main principles, i.e.: of completeness, versatility, explicitness and simplicity of the model itself. The author is aware of the fact that attempts to render sustainable tourism in a model approach had already been made, but it seems that they concerned, in majority, its particular aspects, such as the origins of the phenomenon, its relationships with certain forms of tourist movement or relationships between sustainable development and tourism. However the literature lacked a holistic approach which would take into consideration all most important features of sustainable tourism.

When constructing the theoretical model of sustainable tourism, the author tried to take into consideration the hitherto output of the Polish and international literature, available thanks to the studies of source materials. It enabled the adoption of main model assumptions, and later on, when implementing the deductive method, also the construction of the model itself basing on them. To that end, the author used the form of the mathematical function and notation.

2. SUSTAINABLE TOURISM – A REVIEW OF MAIN IDEAS

The conception of sustainable tourism refers to the wider conception of sustainable development, which stresses the need of rational management of natural environment resources. The first in the global scale sign of the necessity of change in the general conception of economic development was the report of the Secretary-General of the United Nations U Thant entitled 'Man and His Environment', published in 1969. Significant was also the 1st Report of the Club of Rome entitled 'Limits to Growth', published in 1972. The problems of the threat to the natural environment were the main subject of discussion during the UN conference in Stockholm (the so called Stockholm Conference), organised in the same year. At that time, the term 'sustainable development' was introduced. The next milestone in the worldwide discussion on sustainable development was the publication of the report entitled 'Our Common Future', which contained a summary of the activity of the World Commission on Environment and Development (the so called Bruntland Commission). This document adopted the fundamental, still valid, assumption that sustainable development 'seeks to meet the needs and aspirations of the present without compromising the ability to meet those of the future'. In 1992 in Rio de Janeiro the United Nations Conference on Enrivonment and Development (the so called 'Earth Summit') took place. During that conference two documents, significant from the point of view of the sustainable development conception, were adopted. These were the so called Rio Declaration, containing 27 principles defining rights and duties of nations in the field of sustainable development, and AGENDA 21, the global action plan referring to the actions necessary in order to achieve sustainable development and high life quality (Kowalczyk, 2010; Niezgoda, 2006).

Conceptions of tourism development referring to the principles of sustainable development began to appear in the international literature on a larger scale in the mid 1980s. It should however be noted that as early as 1965 W. Hetzer formulated the definition of the so called *responsible tourism*, which in fact was very close to these principles [Blamey, 2001, as cited in Kowalczyk, 2010]. It seems, though, that the moment which began the discussion on new ways of developing tourism was when the conception of the so called *alternative tourism*[1] arose. J. Krippendorfer, who published in the *Annals of Tourism Research* in 1986 the article entitled 'Tourism in the system of industrial society', is considered the author of its definition. As the name itself suggests, it arose in opposition to the so called mass tourism, viewed by the proponents of this conception as the so called 'bad option'. Alternative tourism, often identified with small-scale tourism and treated as the 'good option', was meant to oppose the 'bad option' (Clarke, 1997; Lanfant, Graburn, 1992; Weaver, 2001).

In the same period various conceptions connected with the so called ecotourism began to appear in the international literature. H. Ceballos-Lescuráin (1987) is considered the author of its first definition. At the same time scholars began to introduce terms similar to *ecotourism* or *alternative tourism* such as *green tourism (tourisme vert, nature-based, naturnäher), soft tourism (saufer Tourismus), nature tourism, environmental friendly/environmentally sensible tourism, responsible tourism (angepast), discreet tourism, appriopriate tourism, ecoethnotourism* (Boo, 1990, Cater, Lowman, 1994, Krippendorf et al., 1998; Niezgoda, 2006). It should be noted that the authors of these definitions stressed first of all the (desired) way of cultivating tourism, types of values (mainly natural) and the (small) scale of the phenomenon. They often used the evaluating approach which juxtaposed the 'new' forms of tourism with these 'old', often identified with mass tourism.

A broad overview of diverse definitions of sustainable tourism was included in R.W. Butler's paper entitled 'Sustainable tourism: a state-of-the-art review' (1999). This author, who is skeptic towards views that sustainable tourism constitutes a panaceum for contemporary tourism's problems, presents his own view on its essence. He claims that sustainable tourism can be seen in two ways (Butler, 2005). Firstly, from the semantic- dictionary side, taking into consideration its feature of sustainability as a warranty of long- term survival on the market. According to M. Mika (2008) such an approach seems to be closer to the representatives of the economic party, who stress the problem of self- maintenance of tourism development. The second way of understanding sustainable tourism by Butler is much closer to the conception of sustainable development. It suggests treating sustainable tourism as a tool for the development of reception areas without breaking the principles of sustainable development. As one may guess, this attitude is closer to the representatives of the natural sciences and the humanities. Butler's views on ambiguity in understanding the term sustainable tourism are supported by A. Niezgoda (2006), who claims that conception of sustainable tourism ocurred as a result of research on interrelations between tourism, environment and development. According to this author sustainable tourism is treated by scholars as a tool for realization of sustainable development or a tool for the development of tourism itself.

Totally different scientific basis of sustainable tourism conception (or sustainable development through tourism) is presented by Bryan H. Farell and Louise Twinning-Ward (2003). In the article entitled 'Reconceptualizing Tourism', published in 2003 in the *Annals of Tourism Research*, they postulate a total change in the methodological approach towards the studies of tourism, sustainable tourism included. These authors criticize strongly the hitherto, according to them most wide-spread, way of conducting research in the field of tourism, which is based on narrow specialization, linear reductionism as well as determinism assuming predictability of phenomena and presence of cause and effect. They claim that such an approach, due to complexity and unpredictability of behaviour of tourist systems and systems influencing tourism, cannot guarantee satisfactory results. Instead, they propose a new paradigm that is based on the interdisciplinary approach encompassing relatively new fields, such as: ecosystem ecology, ecological economics, global change science and complexity theory. These authors assume that natural and social systems do function in a relatively independent and non-linear way and therefore postulate implementation of the complex adaptive systems theory into the studies of tourism. Simultaneously, they introduce the notions of comprehensive tourism system and complex adaptive tourism systems – CATS.

Apart from the broad and varied in views discussion on the essence of sustainable tourism present in numerous scientific publications, also institutional documents devoted to sustainable tourism that are of declarative, explanatory or quasi-normative character are winning wide renown (Table 1). Among numerous publications of this type one should note i.a. the Charter for Sustainable Tourism (adopted in 1995), whose signatories agreed that development under the influence of tourism should refer to the principles of sustainable tourism, which meant that it should take into consideration the long-term needs of the natural environment, affect positively a given economy and be accepted in terms of ethics and culture by local communities. The same document claims that tourism should contribute to sustainable tourism through strict integration with the natural and the antropogenic environment on reception areas. Also in 1995 World Travel and Tourism Council, United Nations World Tourism Organization and Earth Council adopted the document entitled 'Agenda 21 for the Travel and Tourism Industry: Towards Environmentally Sustainable Development'. This document defines i.a. the priorities of sustainable tourism. In 1999 the United Nations World Tourism Organization published the Global Code of Ethics for Tourism, which took into consideration the postulates of sustainable tourism. In 2004 the same organization defined the principles of sustainable tourism as those which refer to all forms of tourism (mass tourism included). At the same time, it was highlighted that in order to ensure a long-term balance the principles of sustainable development in tourism must concern environmental, economic and socio- cultural issues to the same degree (Sustainable development of tourism. Conceptual definitions, 2004). Finally, in 2008, during the World Conservation Congress, which took place in Barcelona, the document containing Sustainable Tourism Criteria was adopted.

Table 1. Selected documents concerning sustainable tourism

Document	Publishing subject	Year	Place of publication
Charter for Sustainable Tourism	World Conference on Sustainable Tourism	1995	Lanzarotte, Canary Islands
Agenda 21 for the Travel and Tourism Industry: Towards Environmentally Sustainable Development	WTTC, UNWTO, Earth Council	1995	Madrid (1996)
Berlin Declaration	International Conference of Environment Ministers on Biodiversity and Tourism	1997	Berlin
Global Codes of Ethics for Tourism	UNWTO	1999	Santiago de Chile
The encyclopedia of ecotourism	Weaver D.B. (ed.) CABI Publishing	2001	Oxon (UK) – New York (USA)
Sustainable development of tourism. Conceptual definitions	UNWTO	2004	Madrid
Global Sustainable Tourism Criteria	World Conservation Congress (Rainforest Alliance, UNEP, UNWTO)	2008	Barcelona

To sum up the deliberation concerning the issue of sustainable development of tourism one should repeat, i.a. after the United Nations World Tourism Organization (2004) that sustainable development should be applicable (as much as possible) to all forms of tourism, including mass tourism. And the principles defining sustainable development in tourism should refer to natural, socio-cultural and economic aspects connected with tourism – by striving to achieve the state of balance between them.

3. SELECTED MODELS OF SUSTAINABLE TOURISM

As it was mentioned, sustainable tourism is an area of interest of many scholars, activists and practicians in various countries. It seems, though, that both in the international and in the Polish literature the descriptive method dominates. It puts stress on explaining the conception of sustainable tourism, which is often done from different scientific positions. Apparent is the evolution of views on its essence. In the first period natural aspects were emphasized first of all – in the context of preserving natural environment resources against the threats of tourism. Now, however, we are dealing with the situation in which economic and socio-cultural aspects are seen as well. To a large extent it is thanks to i.a. the publications of the United Nations World Tourism Organization, which emphasized the necessity of striving for balance in fulfiling needs of all tourism stakeholders functioning within natural and socio-economic environment. The change in the approach towards sustainable tourism during past few decades is also expressed in the abandonment of evaluation of various tourism forms according to these criteria. Nowadays, it is stressed that the principles of sustainable tourism should be taken into consideration in all kinds of tourism, including so unpopular among the 'orthodox activists' mass tourism. This evolution, with the consideration of relations between alternative tourism, ecotourism and mass tourism and their relation to sustainable tourism, is presented i.a. by A. Niezgoda (2006). In a simpler form it can be presented graphically as in the Figure 1.

Model relationships between sustainable tourism and unsustainable tourism (often identified with mass tourism) are an area of interest of other authors as well. Among them are, i.a., D.A. Fennel (1999) and D.B. Weaver (1999), who claim that there is no way to designate a clear boundary between sustainable and unsustainable forms of tourism. The former introduces, in relation to various aspects of tourism (attractions, transportation, accommodation, product), kind of degrees (stages) of sustainable tourism. The latter, in turn, claims that mass tourism (closer to unsustainable tourism) constitutes a kind of continuum of alternative tourism (closer to sustainable tourism), so they cannot be treated as separate, opposing categories. These authors' opinions can lead to two kinds of conclusions. On one hand, it is postulated that the principles of sustainable development should be taken into consideration as much as possible in all forms of tourism (Figure 1). In such a case we deal with the desired direction of change from unsustainable tourism to susainable tourism. On the other hand, assuming D.B. Weaver's point of view on mass tourism (more unsustainable) as a continuum of alternative tourism (more sustainable), one can see a more undesirable direction of change from sustainable tourism to unsustainable tourism. Both situations are ilustrated by bilaterally oriented arrows in Figure 2.

A similar conclusion concerning possibilities of occurrence of undesirable direction of change can be drawn after the analysis of three theoretical models of tourism: 1) of tourist area life cycle (TALC) by R.W. Butler (1980); 2) of tourist space by S. Liszewski (1995), and 3) of changes in the natural environment under the influence of tourism by D. Zaręba (2010).

The curve of dependences occurring between the number of tourists on a given reception area and the time (Butler), the level of tourist space transformation (Liszewski), and the level of the environment devastation (Zaręba) is very similar. After the analysis of the curve in each model (after simplification) one can distinguish 4 stages of changes in the direction from the state of the original balance to the state of a new balance – in transformed, i.e. naturally devastated, environment (Figure 3).

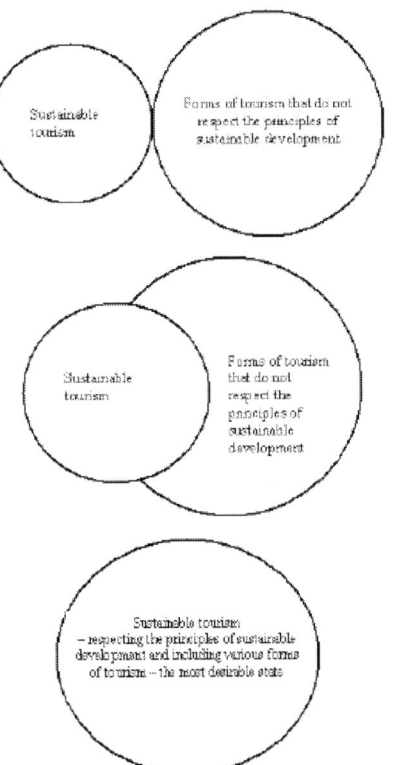

Figure 1. Sustainable tourism and various forms of tourism – the evolution of approach

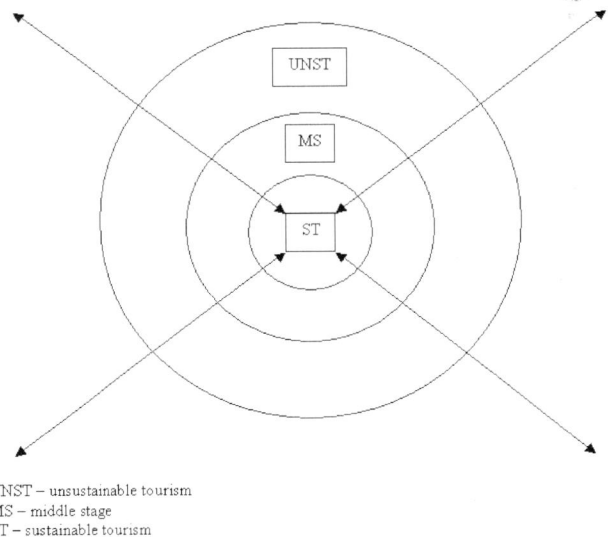

UNST – unsustainable tourism
MS – middle stage
ST – sustainable tourism

Figure 2. Sustainable and unsustainable tourism as a continuum of bi-directional changes

A model conception of diverse degrees (stages) of development (functioning) of sustainable tourism, in relation to different (in terms of environment and socio-economics) reception areas was proposed also by C. Hunter (1997, as cited in Mika, 2008). This author, after a contrastive analysis of the position of tourism and the position of sustainable development within diverse areas, distinguished four variants of functioning of tourism within sustainable development. This conception can be graphically illustrated with a graph of decreasing function that indicates relationships between tourism and sustainable development (Figure 4). Controversy in Hunter's model lies in the fact that it excludes the possibility of a wide-scale tourism development that would take into account the principles of sustainable development. Therefore, this model undermines the idea of sustainable tourism as the one that takes into account the principles of sustainable development.

Polish scholars also made an attempt to present the essence of sustainable tourism in a model form. These were M. Durydiwka, A. Kowalczyk & S. Kulczyk (2010). These authors assumed that the conception of sustainable tourism (ST) concerns mainly three types of tourism, i.e.: 1) related to the natural environment values (STnatural.); 2) related to the cultural environment values (STcultural); 3) requiring from tourists certain skills (STqualifying). Taking into account these types of tourism they presented the idea of sustainable tourism as the following formula:

$$ST = STnatural + STcultural + STqualifying +$$
$$+(STnatural / k \times ST cultural / k \times STqualifying / k)$$

(1)

k – the correction factor.

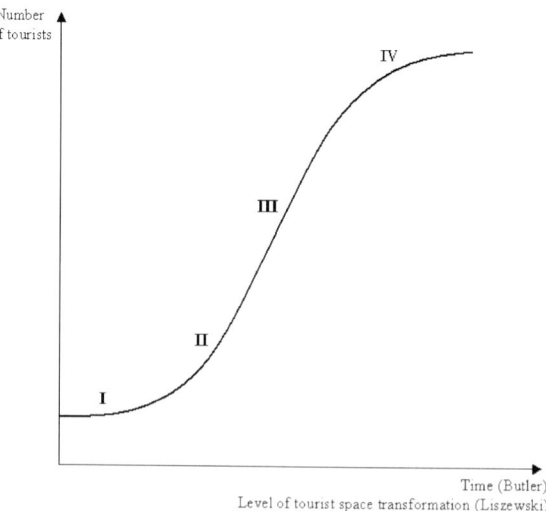

Time (Butler)
Level of tourist space transformation (Liszewski)
Level of natural environment degradation (Zaręba)

Stage \ Conception	Tourist area life cycle by R. Butler (1980)	Changes in the natural environment (based on: D. Zaręba, 2010)	Types of tourist space by S. Liszewski (1995)
I	Exploration	Original balance	Exploration
II	Introduction	Threat	Penetration
III	Development	Degradation	Colonization
IV	Consolidation and stagnation	New balance	Urbanization

Figure 3. Tourism in the function of time, spatial changes, and environmental changes

According to its authors, this formula refers to the holistic conception of sustainable tourism, which means that it should be understood as a combination of various forms of tourism, complemented by common objectives, such as: care for the natural environment, limiting the negative effects for local population, bringing economic benefits to reception areas and meeting the needs of tourists.

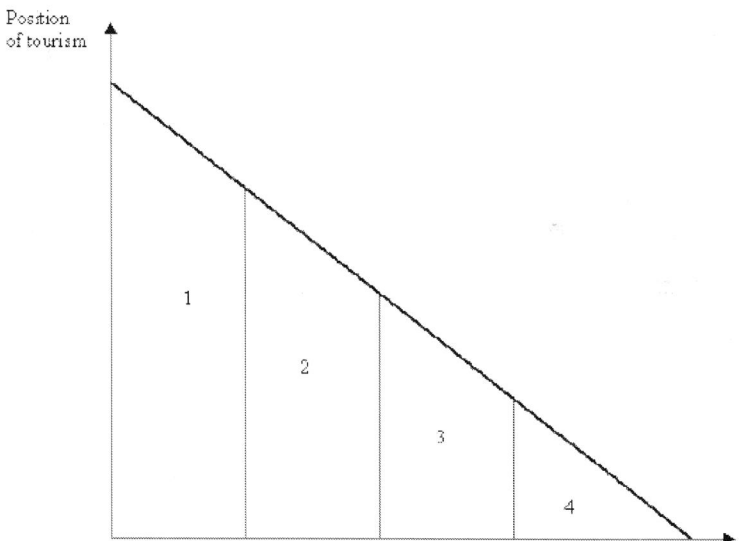

Variant	Position of tourism	Position of sustainable development
1	Domination (imperative) of tourism	Very weak
2	Tourism determined by product	Weak
3	Tourism determined by environmental issues	Strong
4	Minimalised tourism	Very strong

Figure 4. Variants of functioning of tourism in sustainable development

4. THEORETICAL, SHORT-TERM MODEL OF SUSTAINABLE TOURISM

The model is presented in the graphic (Figure 5) and descriptive form, through a presentation of: purposes and conditions of its construction, main assumptions that the model is based on, adopted variables, model factors of balance and its disturbance (notation), factors affecting variables, and possibilities and restrictions on using the model.

4.1 Purposes and conditions of the model's construction

The purpose of the sustainable tourism model construction is to present in a complete, explicit and as simple as possible form the essence of sustainable tourism in the short-term perspective. The author intended the proposed model, designed as a theoretical construct, to render in the most complete way the ideas of sustainable tourism, and at the same time to be appropriate for teaching and guiding purposes as well as to constitute a theoretical basis for detailed application models. The model is intended to be versatile, i.e. to be applicable in all conditions, on every reception area, for every type of tourism. Another condition, which was required in order to meet all the other criteria, was the necessity to use mathematical function dependencies and notation (explicitness of the model). The simplicity of the form, facilitating understanding of the model, is ensured through minimilization of the number of variables and by the graphic illustration of the model. An additional intention of the author was to take into consideration the possibility of occurrance of change of independent variables and their influence on dependent variables (the dynamic factor). It allows to observe, and especially to project the effects of these changes, in the context of their consequence for sustainable tourism.

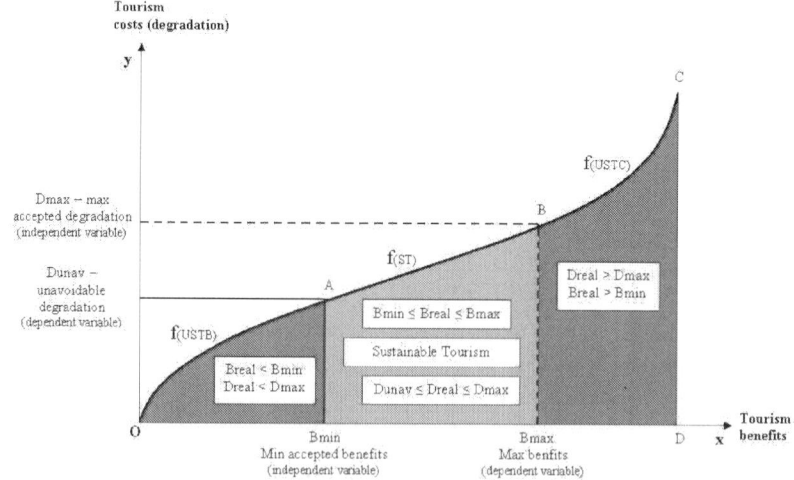

Figure 5. Theoretical (short-term) model of sustainable tourism

Assumptions for the sustainable tourism model construction

1. The assumed objective of sustainable tourism on a given tourist reception area has been the striving for the state of balance in fulfiling needs (reaping benefits) of two main groups of stakeholders, i.e.:

   o   tourists – who visit the tourist reception area in order to fulfil their tourist needs (to reap benefits);

   o   community inhabiting or working in favour of tourism on the reception area (local population, transactors operating tourists, public authorities) – which agrees on or acts in favour of tourism development, because it acknowledges a chance to fulfil its needs (to reap benefits).

At the same time, the accepted level of the degradation of the natural and socio-cultural environments, which includes tourist resources of a given reception area (in the wide sense of tourist potential), cannot be exceeded.

It has also been assumed that the increase in (short-term) benefits reapt by tourists and the inhabitants of the areas that they visit – related to developing tourism – results in (in principle) the increase in the level of the degradation of the natural and socio-cultural environments. In this context, the degradation can be treated as a kind of an unavoidable environmental cost that must be borne in connection with developing tourism. This assumption indicates the short-term perspective of functioning of the model. For, it is obvious that in a long-term perspective, after exceeding the accepted level of degradation it will not be possible to reap further benefits, at the expense of already devastated environment.

1. The author has also assumed an auxiliary assumption concerning the possibility of occurrence of reverse dependency between the benefits reapt by tourists and the benefits reapt by the local community (presented in the graph as a decreasing function), which in sustainable tourism results in the necessity to seek an 'area' of balance in fulfiling the needs of both groups of stakeholders (auxiliary model – Figure 6).

2. The author has also assumed an auxiliary assumption concerning the possibility of occurrence of reverse dependency between the benefits reapt by tourists and the benefits reapt by the local community (presented in the graph as a decreasing function), which in sustainable tourism results in the necessity to seek an 'area' of balance in fulfiling the needs of both groups of stakeholders (auxiliary model – Figure 6).

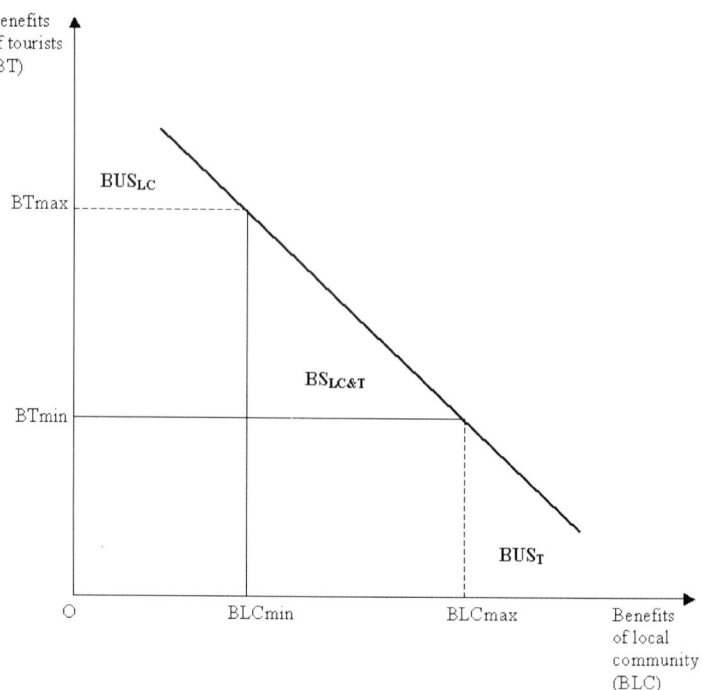

Figure 6. Auxiliary model – the benefits of tourists and the benefits of the local community in sustainable tourism

$$f(x)\!: y = ax + b;\ a{<}0,\ x{>}0,\ y{>}0$$

where:

BLCmin – minimal benefits of the local community (independent variable)

BTmin – minimal benefits of tourists (independent variable)

BLCmax – maximal benefits of the local community (dependent variable)

BTmax – maximal benefits of tourists (dependent variable)

$BS_{LC\&T}$ – sustainability between the benefits of the local community and the benefits of tourists

BUS_{LC} – unsustainability of the benefits of the local community

BUS_{T} – unsustainability of the benefits of tourists

BLC_{max} – maximal benefits of the local community (dependent variable)

BT_{max} – maximal benefits of tourists (dependent variable)

4.3 Explanations for the main model

1. Benefits from tourism – benefits reapt by tourists visiting a given reception area and benefits of the local population (including transactors, public authorities and other organizations), resulting from development of tourism:

   o min accepted benefits (Bmin): denotes the minimal accepted level of fulfiling needs of tourists and local population, beneath which the reapt benefits will be evaluated as insufficient; its size is measured with the numerical value of the Bmin point on the Ox axis of the model graph;

   o max benfits (Bmax): denotes the maximal accepted (in sustainable tourism conditions) level of fulfiling needs of both tourists and local population; its size is measured with the numerical value of the Bmax point in the Ox axis of the model graph;

   o real benefits (Breal): the real level of benefits reapt by tourists and local community in relation to tourism developing on a given area.

1a. In the component of benefits there are two basic groups of participants (tourists, local community), which can have opposing interests. In order to take into account the level of balance (sustainability) between the benefits of tourists and the benefits of the local community, as an element of general balance (sustainability), the author has produced an auxiliary model of partitive balance (sustainability) in the benefit component (Figure 6). The assumptions of this model have been transferred to the Ox axis of the main model.

2. Costs of tourism development – degradation of the natural and antropogenic (social, cultural, economic) environments on a tourist reception area, resulting from developing tourism:

   o max accepted degradation (Dmax): denotes the highest accepted in sustainable tourism (i.e. not resulting in irreversible changes) level of degradation of both environments; its size is measured with the numerical value of the Dmax point on the Oy axis of the model graph;

   o unavoidable degradation (Dunav): denotes the level of unavoidable degradation of both environments resulting from developing toursim; its size is measured with the numerical value of the Dunav point on the Oy axis of the model graph;

   o real degradation (Dreal): the real level of degradation of the natural and antropogenic environments occurring on a reception area in relation to tourism developing there.

Independent and dependent variables used in the model

In the model there are two pairs of interelated independent and dependent variables.

Table 2. Independent and dependent variables in the model of the sustainable tourism.

Independent variables	Dependent variables
Min accepted benefits (Bmin)	Unavoidable degradation (Dunav)
Max accepted degradation (Dmax)	Max benfits (Bmax)

1. Min accepted benefits (Bmin – independent variable) reapt by tourists and the community that hosts them; they result in certain unavoidable level of degradation (Dunav – dependent variable) of the natural and antropogenic environments on an analysed tourist reception area.

2. Max accepted degradation (Dmax – independent variable) of both environments denotes the max level of benfits (Bmax – dependent variable) which can be reapt by tourists and the local population in sustainable tourism, i.e. without causing irreversible environmental changes.

4.4 Conditions for sustainable tourism

Table 3. Model conditions for sustainable tourism

Sustainable tourism	General conditions	
	Component of environment	Component of benefits
	$\lvert Dmax \rvert - \lvert Dunav \rvert \geq 0$	$\lvert Bmax \rvert - \lvert Bmin \rvert \geq 0$
	$\lvert Dunav \rvert \leq \lvert Dreal \rvert \leq \lvert Dmax \rvert$	$\lvert Bmin_n \rvert \leq \lvert Breal \rvert \leq \lvert Bmax \rvert$
Sustainable tourism by components	Minimal conditions	
	Component of environment	Component of benefits
	$\lvert Dmax \rvert - \lvert Dunav \rvert = 0;$	$\lvert Bmax \rvert - \lvert Bmin \rvert = 0;$
	but: $\lvert Dmax \rvert > 0 \;/\backslash\; \lvert Dunav \rvert > 0$	but: $\lvert Bmax \rvert > 0 \;/\backslash\; \lvert Bmin \rvert > 0$
	$\lvert Dunav \rvert = \lvert Dreal \rvert = \lvert Dmax \rvert$	$\lvert Bmin \rvert = \lvert Breal \rvert = \lvert Bmax \rvert$
Function $f_{(ST)}$ – describing the existence of sustainable tourism for both components	General condition	
	$f_{(ST)} = \{x: x \in [\,\lvert Bmin \rvert , \lvert Bmax \rvert\,]; \lvert Bmax \rvert - \lvert Bmin \rvert \geq 0\}$	
	Minimal condition	
	$f_{(ST)} = \{x: x = \lvert Bmin \rvert ; \lvert Bmax \rvert - \lvert Bmin \rvert = 0\}$	

4.5 Model disruption of sustainability

Table 4. Model disruption of sustainability.

Type of disruption	Condition	Description
Lack of balance (unsustainability) in the component of benefits, balance (sustainability) in the component of environment.	$\|Breal\| < \|Bmin\|$; $\|Dreal\| < \|Dmax\|$	It occurs when the real benefits (Breal) are smaller than the minimal benefits (Bmin). At the same time the level of real degradation (Dreal) is lower than the level of accepted degradation (Dmax).
Function $f_{(USTB)}$ – describing the lack of balance (unsustainability) in the component of benefits while maintaining balance (sustainability) in the component of environment.	$f_{(USTB)} = \{x: x \in [O, \|Bmin\|]\}$	

Type of disruption	Condition	Description
Lack of balance (unsustainability) in the component of environment while maintaining of balance (sustainability) in the component of benefits.	$\|Dreal\| > \|Dmax\|$, $\|Breal\| > \|Bmin\|$	It occurs when the real degradation (Dreal) is bigger than the accepted degradation (Dmax). At the same time the real benefits (Breal) are bigger than the minimal benefits (Bmin).
Function $f_{(USTC)}$ – describing the lack of balance (unsustainability) in the component of environment while maintaining balance (sustainability) in the component of benefits.	$f_{(USTC)} = \{x: x \in [\|Bmax\|, \|D\|]\}$	

Factors affecting independent variables, as determinants of sustainable tourism

1. The accepted level of degradation (understood as the highest accepted in sustainable tourism, i.e. not causing irreversible changes, level of degradation of the natural and antropogenic environments) depends on the type of ecosystem and features of the social enrivonment occurring on a tourist reception area. In the natural component low level of accepted degradation is characteristic for natural and close to natural ecosystems that are very vurnerable to external stimuli. In turn, higher level of accepted degradation is characteristic for significantly transformed ecosystems which are not carriers of special natural values. In the antropogenic component, the most vulnerable to degradation will be close, traditional communities that do not maintain lively contacts with the outer world. In such a case, in order to fulfil the sustainable tourism condition, the accepted degradation level should be as low as possible.

2. The expected minimal level of benefits (taking into account the assumptions of the auxiliary model – Figure 6) that both gropus of

tourism stakeholders (tourists and local population) expect to reap on a given reception area depends on their expectations of tourism. Although, the lowest accepted level of benefits reapt by permanent residents will depend on the features of that community, such as: age structure, education level, environmental and cultural awareness, system of values, self-esteem, hitherto quality of life, professional activity, expectations of development of local tourist economy, local authorities and elite activity. As far as tourists are considered, the case is similar. The level of minimal benefits that they expect will depend on socio-cultural features of that collectivity. They will constitute the basis for the tourists' subjective assessment of the local tourist product (including, i.a. values, tourist management, prices). This product will have to meet the needs of tourists enough for the tourists to think that for the price they are ready to pay they will get the minimal accepted level of benefits related to tourist trip to that location.

3.   Taking into account the above-mentioned model assumptions, the sustainable tourism area – presented on the graph as:

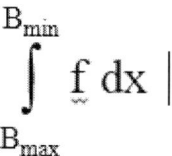

$$\left. \int_{B_{max}}^{B_{min}} \underset{\sim}{f}\, dx\ \right|$$

will depend on: 1) the resistance of the natural and antropogenic environments to the negative influence of tourism, denoted with the location of the Dmax point on the Oy axis of the model graph (independent variable); and 2) the minimal accepted level of benefits that local population and tourists expect to reap, denoted with the location of the Bmin point on the Ox axis of the graph (independent variable). The model tourism sustainable area will depend on one hand on the willingness of both groups of stakeholders to resign from the short-term benefits that they want to reap from tourism (possibly small numerical value of the Bmin point on the Ox axis), on the other hand on the features of the environment that determine its vulnerability to degradation by tourism (possibly high numerical value of the Dmax point on the Oy axis).

Implementation of the model – Possibilities and limitations

1.   The implementation of the model for the scientific-educational (explanatory) purposes – the model can be used in order to explain the essence and the principles of sustainable tourism, and especially to determine the interrelations occurring between all the stakeholders of tourism and the natural and antropogenic environments in which tourism is being developed. The construction of the model enables analyses of these interrelations in dynamic hold, which reveals

consequences for all the tourism stakeholders induced by a change of independent variables used in the model. Another advantage of the model is its versatility, i.e. the fact that it is applicable in relation to all types of tourism (tourist movement) and reception areas. For, in every situation the same factors (determinants), occurring in the model as independent variables determining framework for the development of sustainable tourism, are taken into account. Also, in all analysed cases the assumed model conditions must be fulfilled. Versatility and explicitness of the model manifest themselves also in the utilization of the graphic way of presenting function dependencies and notation that together define the main assumptions, interrelations and conditions included in the model.

2. The implementation of the model for the purposes of application – the model can be used in order to find out to what extent will the development of various types of tourism on a given reception area fulfil the principles of sustainable tourism. Particular types of tourism should be analysed both in terms of demand, as a form of tourist movement, and in terms of supply, as corresponding types of tourist products (in the widest sense of this term). Practically, one should make an attempt to construct individual models for each type of tourism. This will be possible after choosing appropriate measures (indices) determining the values of particular variables. After building individual model graphs one will be able to compare the obtained ranges (size) of sustainable tourism, characteristic for particular types of tourism. Results of such an analysis may be especially useful in order to determine the types of tourism preferable for a given area – taking into accaount the conditions of sustainable tourism.

It seems that the main barrier affecting negatively the application type of implementation of the proposed model of sustainable tourism can be difficulties related to the quantification of the adopted variables in detailed models. It would be easiest to express time in financial values, but this may not always be possible and appropriate. It is also possible to use other indices published i.a. in the publications of the United Nations World Tourism Organization and other organizations (programmes), such as 'Making Tourism More Sustainable' (2005 as cited in Kowalczyk, 2010; *The VISIT initiative*, 2004). When selecting indices one should make sure that they fulfil the criteria for the ideal index of sustainable development as much as possible. These criteria are: 1) simplicity of identification and measurement, 2) natural and/or social, cultural, economic, political significance, 3) stability, 4) simplicity and low cost of measurement, 5) sensitivity and quickness of reaction to changes, 6) intelligibility and explicitness (based on Hughes, 2002, as cited in Kowalczyk, 2010).

At the same time, one should not forget that the variables used in the main model are internally diverse. One group includes netto benefits reapt by both tourists and local population (including transactors), while the other concerns total environmental costs manifesting themselves in the degradation of the natural and antropological environments. In order to determine values of these variables one should consider each of their elements individually and assume an appropriate breakpoint (e.g. according to the assumptions made in the auxiliary model concerning the component of benefits – Figure 6.) One can also consider the solution of application simplication of the entire model. In such a case only one (breakpoint) component of a given variable would be taken to quantification. E.g. for the independent variable 'required benefits' such an operation would include defining the minimal accepted level of benefits reapt by local population and then treating it as the assumed breakpoint level (with the underlying assumption that sole appearance of tourists on a given area testifies of the fact that tourists reap their accepted level of benefits.) The same operation can be used while dealing with the other variable, making the choice of its component dependent of the type of reception area (for sure, for the areas naturally valuable it should be the maximal, accepted for given ecosystems, level of natural environment degradation.)

Another significant problem in the practical implementation of the model is to find the appropriate functional interrelation between assumed variables (costs vs. benefits) in the detailed models, both for breakpoint and for intermediate values, which will decide what the function of sustainable tourism for a given type of tourism on a given reception area will look like. In the main model only the general rule of interrelation between costs and tourist benefits (presented in the graph as an increasing function) was taken into account. It is the ability to determine the shape of the curve through defining the values of variables (breakpoint and intermediate) for various types of tourist movement on a given reception area that will allow the use of the assumptions of the main model on a wider scale for the purpose of application.

Both above-mentioned problems (quantification of variables and finding functional interrelations between them) are important in terms of the application use of the presented model, since they directly affect the ability to define the model size of sustainable tourism.

SUMMARY

Because of the fact that the sustainable tourism literature is dominated by the descriptive style and because it is multidirectional, which leads to ambiguities in defining the phenomenon, the author has made an attempt to construct a theoretical model of sustainable tourism which would render both its essence and main features.

The article presents the theoretical, short-term model of sustainable tourism. It has been designed basing on the adopted assumptions that define the essence of sustainable tourism. They concern striving for the state of balance between the needs of tourists and the needs of local community, while maintaining the values of the natural and socio-cultural environments that occur on reception areas. In other words, the article means that kind of tourism which is satisfactory for tourists and the local population (including transactors working in favour of tourism) and which does not cause irreversible degradation of the natural and antropogenic environments.

The model is intended to fulfil the criteria of completeness, versatility, explicitness and simplicity. To that and, the author has used the graphic form of mathematical function and notation. The model constructed in this way can be implemented for explanatory- educational purposes as well as application purposes (after selecting appropriate indices). The author hopes that the theoretical model of sustainable tourism presented in this article will constitute a complement to the output of the studies of sustainable development in tourism, especially in their theoretical aspects.

NOTES

A. Niezgoda [2006] claims that the conception of alternative tourism stems from the so called Hippie contrculture, which arose in the 1960s in the USA and later spread in Europe. In this context, alternative tourism was meant to be the new way of travelling that would not destroy the environment and authentic relations between people.

REFERENCES

1. *Agenda 21* (1992). UN Conference on Environment and Development, Rio de Janeiro
2. *Agenda 21 for the Travel and Tourism Industry: Towards Environmentally Sustainable Development* (1996). WTTC, UNWTO, Earth Council
3. *Berlin Declaration* (1997). International Conference of Environment Ministers on Biodiversity and Tourism, Berlin
4. Blamey R.K. (2001). *Principles of ecotourism*. In: *The encyclopedia of ecotourism*, Weaver (ed.), pp. 5–22, CABI Publishing, , Oxon– New York
5. Boo E. (1990). *Ecotourism: the potentials and pitfalls, Country case studies*, World Wildlife Fund, Washington
6. Butler R.W. (1980). The Concept of a Tourist Area Cycle of Evolution: Implication for the Managment of Resources, *The Canadian Geographer*, 25, pp. 151–170
7. Butler R.W. (1999). Sustainable tourism: a state-of-the-art review, *Tourism Geographies*, 1, 1, pp. 7–25
8. Butler R.W. (2005). Problemy miejsc recepcji turystycznej ze zrównoważonym rozwojem, In: *Turystyka w badaniach naukowych*, Winiarski, Alejziak, (ed.) pp. 35–48, AWF w Krakowie, Wyższa Szkoła Informatyki i Zarządzania w Rzeszowie
9. Cater E., Lowman G. (ed.), (1994) *Ecotourism: a sustainable option?*, Published in association with the Royal Geographical Society, Wiley, Chichester, New York

10. Ceballos-Lescuráin H. (1987). The future of ecotourism, *Mexico Journal*, January 17, pp. 13–14 *Charter for Sustainable Tourism* (1995). World Conference on Sustainable Tourism, Lanzarotte Clarke J. (1997). A framework of approaches to sustainable tourism, *Journal of Sustainable* a. *Tourism*, 12, 6, pp. 224–233

11. Durydiwka M., Kowalczyk A. & Kulczyk S. (2010). Definicja i zakres pojęcia 'turystyka zrównoważona', In: *Turystyka zrównoważona*, Kowalczyk (ed.), pp. 21–43, Wydawnictwo Naukowe PWN, Warszawa

12. Farrel B.H., Twinning-Ward L. (2004). Reconceptualization Tourism, *Annals of Tourism Research*, 31, 2, pp. 274–295

13. Fennel D.A. (1999). *Ecotourism. An Introduction*, Routledge, London–New York

14. *Global Codes of Ethics for Tourism* (1999). UNWTO, Santiago de Chile

15. *Global Sustainable Tourism Criteria* (2008). UNEP, UNWTO, Rainforest Alliance, Barcelona Hughes G. (2002). Environmental indicators, *Annals of Tourism Research*, 29, 2, pp. 457–477 Hunter C. (1997). Sustainable tourism as an adaptive paradigm, *Annals of Tourism Research*, a. 24, 4, pp. 850–867

16. Kowalczyk A. (ed.), (2010). *Turystyka zrównoważona*, Wydawnictwo Naukowe PWN, Warszawa

17. Krippendorf J. (1986). Tourism in the system of industrial society, *Annals of Tourism Research*, 13, 4, pp. 517–532

18. Krippendorf J., Zimmer P. & Glauber H. (1988). *Für einen anderen Tourismus, Problemen- Perspectiven-Ratschlage*, Fischer Verl., Frankfurt am Main

19. Lanfant M.-F., Graburn N. (1992). International tourism reconsidered: the principles of the alternative, In: *Tourism altenatives: potentials and problems in the development of tourism*, Smith, Eadington (ed.), University of Pennsylvania Press, Philadelphia, pp. 88–112

20. Liszewski S. (1995). Przestrzeń turystyczna, *Turyzm*, 5, 2. Uniwersytet Łódzki

21. *Making tourism more sustainable. A guide for policy makers.* (2005). United Nations Environment Programme, World Tourism Organisation

22. Meadows D.H., Meadows D.L., Randers J. & Behrens W.W. III. (1972). *Limits to Growth*, Universe Books

23. Mika M. (2008). Sposoby ograniczenia negatywnego wpływu turystyki, In: *Turystyka*, Kurek (ed.), pp. 471–482, Wydawnictwo Naukowe PWN, Warszawa

24. *Nasza wspólna przyszłość* (1991). Raport Światowej Komisji ds. Środowiska i Rozwoju, PWE, Warszawa

25. Niezgoda A. (2006). *Obszar recepcji turystycznej w warunkach rozwoju zrównoważonego*, Wydawnictwo Akademii Ekonomicznej w Poznaniu

26. *Our Common Future* (1987). United Nations Conference on the Human Environment

27. *Rio Declaration on Environment and Development* (1992). UN Conference on Environment and Development, Rio de Janeiro

28. *Sustainable development of tourism. Conceptual definitions* (2004). UNWTO, Madrid Thant U. (1969). *Man and his environment*, UN

29. *The VISIT initiative. Tourism eco-labelling in Europe – moving the market towards sustainability. a.* (2004). ECEAT, ECOTRANS

30. Weaver D.B. (2001). Ecotourism in the context of other tourism types, In: *The encyclopedia of ecotourism*, Weaver (ed.), CABI Publishing, Oxon–New York, pp. 73–83

31. Zaręba D. (2010). *Ekoturystyka*, Wydawnictwo Naukowe PWN, Warszawa

CHAPTER 2

Harmonious Tourism Environment and Tourists Perception: An Empirical Study of Mountain-Type World Cultural Heritage Sites in China

Zhiyong Fan[*], Sheng Zhong, Wei Zhang

School of Economics and Management, Wuhan University, Wuhan, China.

ABSTRACT

This study tested the structural equation model between harmonious tourism environment perception and tourists' loyalty. In this paper, a model is erected to indicate the relationship between harmonious tourism environment perception and tourists' loyalty. With 377 valid questionnaires from 8 mountain-type world cultural heritage sites in China, the author tested the model by SPSS and AMOS. It was found that the perception of natural ecosystem was the most influential factor affecting tourists' entire environment perception, social relations system was the second place and cultural ecosystem was the third place. The entire harmonious tourism environment perception also greatly determines tourists' loyalty. At last, some suggestions were given to make tourism destination develop sustainably based on this study.

Keywords: Harmonious Tourism Environment; Tourist Perception; World Cultural Heritage

1. INTRODUCTION

Tourist perception is the tourist's comprehensive perception on the attractions, environment, products and services in tourism destination. Tourist perception and behavior is the hot spot in tourism research, and the contemporary literatures mainly focus on the theory model of "quality-value-satisfaction-loyalty", which has been fully or partly tested through different cases, variables in many studies [1-5]. However, there is still lacking the empirical research on the tourist's all-round perception on the destination's environment, as well as the relationship among perception, satisfaction and loyalty.

Tourism environment is the foundation of all the tourism attractions, products and activities. In the current process of tourism development, harmonious tourism environment represents the future of tourism environment construction. As a necessary route to sustainable development in destination, harmonious tourism environment is a harmonious state of natural ecosystem, cultural ecosystem and social relation system in tourism destination under the direction of scientific development perspective [6]. Till the end of 2009, there has been 8 mountain-type world cultural heritage sites (hereinafter "8 mountains"), including Mount Taishan, Mount Huangshan, Mount Wudang, Mount Lushan, Mount Emei, Mount Wuyi, Mount Qingcheng and Mount Wutai, all of which also sever as the most famous tourism destinations in China. Zhang et al. [7] discussed the fundamental theory model of the tourism harmonious environment's construction in mountain-type cultural heritages. Based on Zhang's research, this paper takes an empirical approach to study the relationship between the perception of tourism harmonious environment and tourists' loyalty from the perspective of tourist perception.

2. PERCEPTION OF HARMONIOUS TOURISM ENVIRONMENT

2.1. Harmonious Relationship in Natural Ecosystem

Natural ecosystem is the foundation of all the tourism actions in mountain-type world cultural heritages and serves as a basic component in harmonious tourism environment. In China, all the mountain-type cultural heritage sites boast unique geographical features, beautiful natural sceneries and abundant biological resources. There are 4 mountains in "8 mountains" are both cultural and natural heritages (Mount Taishan, Mount Huangshan, Mount Wuyi and Mount Emei), as well as 2 cultural landscape heritages which represent the perfect combination of human and nature (Mount Lushan and Mount Wutai). Meanwhile, because of the features of instability, sensibility and fragility, mountain-type heritages are very vulnerable under the condition of the mass tourists' pouring. So, this study takes 7 variables, including climate, geology, water, atmosphere, sound, biology and sanitation, as the measuring objects to scale the tourists' perception on the harmonious relationships in natural ecosystem, at the same time provides basis for the natural environment's management and monitoring in tourism destinations.

2.2. Harmonious Relationship in Cultural Ecosystem

Culture plays an essential role in world cultural heritage sites, and cultural ecosystem—the harmonious relationship between human and cultural heritage is the core in mountain-type cultural heritage's environment. The 8 mountain-type cultural heritages in China play a very important role in Chinese culture history. For example, Mount Emei, Mount Qingcheng, Mount Wudang and Mount Wutai are sacred mountains in Buddhism or Taoism; Mount Taishan also served as a sacred site for ancient Chinese emperors to sacrifice to heaven; Mount Wuyi is the origin of Zhuxi culture, known as one subsidiary of Confucianism; Mount Huangshan and Mount Lushan bear many thoughts in Chinese traditional aesthetics. Therefore, the 6 indicators in this part are not only antiquities, architectural complexes, relics and cultural landscapes, which represent the authenticity and integrity of cultural heritage, but also the inheriting state of cultural traditions, as well as the coordination between tourism facilities and environment. By measuring these indicators, this part aims to observe the tourists' perception on the harmonious relationship of human and cultural heritage, and help to the cultural heritage's protection and cultural atmosphere.

2.3. Harmonious Relationship in Social Relation System

Natural and cultural environment cannot be isolated in cultural heritage sites, especially in the process of tourism developing, because they are enormously affected by all kinds of stakeholders, such as the attitudes of local residents, the relationship between residents and operators, the cultural shock and interference taken by tourists. So, the social relation system—harmonious relationship among stakeholders also should be regarded as an important part in the destination's harmonious environment. It is significant to research the relationships in or among the tourists, residents and operators, because since 1970s, with millions tourists pouring in annually, tourism developed in an incredible speed in the "8 mountains" in China. Therefore, from the perspective of tourists, this paper takes the relation between residents and operatorsrelation between hosts and guests, tourism services, tourism commodities, public security and tourists' density, totally 6 indicators, to watch the harmonious condition of social relation system, so as to help refine the management institution and construct harmonious relations among stakeholders.

2.4. Entire Perception on Harmonious Tourism Environment

The perception of harmonious tourism environment is the tourists' entire recognition on destination's environment, constituted by two aspects, one is the satisfaction as well as the other is the participation degree in environmental construction. So, two items about this part were put forward in the questionnaire are: "I am satisfied with the tourism environment" and "I will contribute my effort to tourism environment", respectively represent the tourists' satisfaction and participation.

2.5. Loyalty

A competitive tourist destination is that one can satisfy tourists, and attract more tourists by means of present tourists' revisit and recommendation, so as to attain a stable market. The conception of tourist loyalty stems from customer loyalty, as Opperman [8] argued that, the tourist loyalty on one destination can be comprehended in two aspects, one is the behavior, loyalty is closely related with the frequency of revisit; the other is the manner, loyalty is presented in tourists' revisit intention and recommendation intention. Thus, two items in this part are: "I will come here again" and "I will recommend others to travel here", respectively represent the revisit intention and recommendation intention.

2.6. Theoretical Model and Hypothesis

Based on the analysis of harmonious relationships in natural ecosystem, cultural ecosystem, social relation system, entire perception on harmonious environment and tourists' loyalty, this paper erects a harmonious environment perception theoretical model, including 5 structure variables and 23 items, to measure the effect of natural ecosystem, cultural ecosystem and social relation system to entire perception of harmonious tourism environment, as well as entire perception to tourists' loyalty respectively, as shown in **Figure 1**. At the same time, there are 4 hypotheses in this paper:

H1: Natural ecosystem directly affects the entire perception of harmonious tourism environment.

H2: Cultural ecosystem directly affects the entire perception of harmonious tourism environment.

H3: Social relation system directly affects the entire perception of harmonious tourism environment.

H4: Entire perception of harmonious tourism environment directly affects the tourists' loyalty.

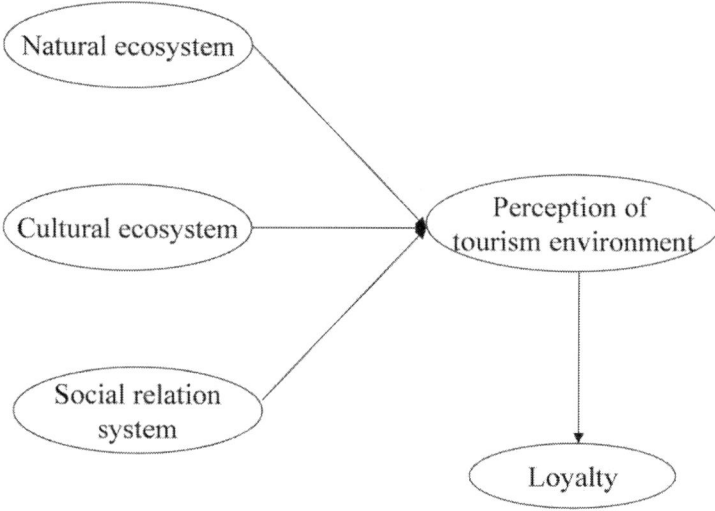

Figure 1. Harmonious tourism environment perception model.

3. RESEARCH METHODS AND QUESTIONNAIRE

3.1. Research Methods

This research utilizes SPSS 13.0 as the tool to conduct descriptive analysis, factor analysis and reliability analysis. Then AMOS 7.0 is used to depict the structural equation model (SEM) of the harmonious tourism environment perception theoretical model, as well as calculate and test the path coefficients. SEM now is broadly used in the realms of psychology, pedagogy, statistics and marketing, and is also more and more applied in tourism research. SEM can test and estimate the casual relations in theoretical model, so it is available in this research.

3.2. Questionnaire Design

The questionnaire is divided into two parts, part one is the 23 items or questions, which are measured based on a five point Likert scale ranging from 1 to 5, where 1 = Strongly disagree; 3 = Neutral; 5 = Strongly agree. Part two is the sample's demographic characters, including gender, age, occupation, education level, as well as tourism pattern. The online questionnaire also includes two questions of timing and visiting time. Timing is the time from responder begins the questionnaire to complete it and is automatically conducted by computer. Visiting time is when the responder's last visiting to "8 mountains".

3.3. Survey Process

The authors have made lots of investigations in "8 mountains" before this study, visited Mount Wudang in June 2008, Mount Lushan in October 2008, Mount Taishan in June 2009, Mount Emei and Qingcheng in October 2009, Mount Wuyi in November 2009, Mount Huangshan in April 2010. In every investigation, the authors not only observed the tourism environment in "8 mountains", but also made a lot of questionnaire surveys and interviews on tourists and local residents, which served as the original data for this study.

The questionnaire survey for this paper is divided into two parts, one is onsite and the other is online. The onsite survey was carried out in Mount Taishan in June 2009, 158 questionnaires were taken back and 154 were identified valid, the percent of validity is 97.5%. With the help of the online survey tool "questionnaire star" (www.sojump.com), the online questionnaires were published from November 20 to December 20, 2009. With the average filling time is 277 seconds, 465 persons who formerly visited "8 mountains" answered the questionnaires and 223 were identified valid after deleting the visiting time before 2008, the percent of validity is 50.1%. For this paper, the authors totally took back 623 questionnaires, valid is 377, and the total validity percent is 60.5%.

With the fast growing of internet, online questionnaire survey is also becoming more and more popular. The responders can get enough time to complete the questionnaire online; at the same time every question is required to answer, so as to avoid the hurry answering and miss answering, both of which frequently occurred in onsite questionnaires. Fang et al. [9] proved the validity of online questionnaire through an empirical research, and in tourism research realm, Chen and Huang [10] studied the influencing factors of tourists revisit decision.

4. RESEARCH RESULTS

4.1. Demographic Analysis

Demographic characteristics are calculated from the samples (see **Table 1**), and female is more than male, most people's age are between 15 - 34 (56.3%), the majority of the samples (78.5%) have the education level of the college, students are the most frequent occupation (48.05) as well as travel with friends is the most popular tourism pattern (41.1%).

Table 1. Sample characteristics.

Items	Options	Frequency
Gender	Male	39.8%
	Female	60.2%
Age	Less than 14	0.5%
	15 - 24	56.5%
	25 - 34	27.3%
	35 - 44	10.6%
	45 - 54	3.7%
	More than 55	0.8%
Education level	Junior high school or below	0.8%
	Senior high school	6.9%
	Undergraduate	78.5%
	Postgraduate	13.8%
Occupation	Enterprise staff	21.5%
	Individual business	5.6%
	Peasants	0.5%
	Students	48.0%
	Liberal professions	6.4%
	Government staff	11.7%
	Others	6.4%
	Government staff	11.7%
	Others	6.4%
Tourism pattern	Join travel agency	17.0%
	Cooperation arrange	6.9%
	Conference	4.2%
	With friends	41.1%
	With family	15.9%
	individual	14.9%

Table 2. Results of descriptive analysis.

Items	Loadings	Reliability	Mean	S.D.
Natural ecosystem				
X1 Climate	0.823		4.11	0.77
X2 Geology	0.809		4.11	0.77
X3 Water	0.781		3.95	0.90
X4 Atmosphere	0.646	0.868	4.02	0.81
X5 Sound	0.636		3.97	0.83
X6 Biology diversity	0.634		3.93	0.84
X7 Sanitation	0.601		3.77	0.97
Cultural ecosystem				
X8 Antiquities	0.842		4.05	0.79
X9 Architectural complexes	0.830		4.03	0.78
X10 Relics	0.787		4.09	0.76
X11 Cultural landscapes	0.787	0.852	4.05	0.82
X12 Inheriting state of cultural traditions	0.703		3.79	0.86
X13 Coordination between tourism facilities and environment	0.540		3.74	0.83
Social relation system				
X14 Relation between residents and operators	0.826		3.68	0.91
X15 Relation between hosts and guests	0.717		3.62	0.92
X16 Tourism services	0.704	0.887	3.57	0.96
X17 Tourism commodities	0.694		3.29	1.09
X18 Public security	0.651		3.80	0.83
X19 Tourists' density	0.534		3.48	0.99
Harmonious tourism environment perception				
Y1 Tourists' satisfaction on environment			4.03	0.73
Y2 Tourists' participation intention to environment protection			3.84	0.96
Tourists' loyalty				
Y3 Revisit intention			3.75	0.94
Y4 Recommendation intention			4.02	0.82

4.2. Factor Analysis

With the help of SPSS 13.0, we can see the descriptive analysis results of 23 items in **Table 2**, with KMO = 0.939, Bartlett test p = 0.000. After rotation, 19 items about tourism environment were deduced into 3 factors, together accounting for 63.244% of the total variance, and all the items' factor loading are more than 0.50. The scores of 2 items in harmonious tourism environment perception, respectively about total satisfaction and participation intention, are relatively high, which shows the tourists' high recognition on tourism environment. In the 2 items of tourists' loyalty, tourists' recommendation inten tion is higher than revisit intention, so the tourists are more likely to recommend others than travel there once more after one visit to "8 mountains". The average score of 7 items in natural ecosystem are the highest in 5 structure variables, only the sanitation is relatively low. Regarding the 6 items in cultural ecosystem, the scores of antiquities, architectural complexes, relics and landscapes are higher than 4, yet the scores of inheriting state of cultural traditions and coordination between tourism facilities and environment are less than 4. However, all the scores of 6 items in social relation system are lower than 4, especially tourism commodities' score is the lowest 3.29, which indicates the tourists' low satisfaction on social relation system. Finally, with the reliabilities (α Cronbach) of 3 factors more than 0.80, so it is appropriate for structural equation model (SEM) analysis in this study.

4.3. Structure Equation Model (SEM) Analysis

With the help of AMOS 7.0, **Figure 2** shows the estimated path coefficients of the harmonious tourism environment perception model and all of the 4 hypotheses are accepted.

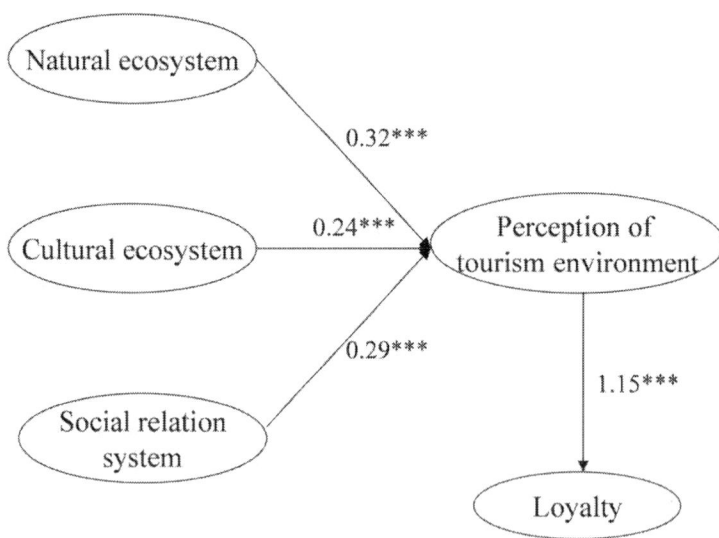

Figure 2. Structural equation model test. (Notes: ***indicates p < 0.001, two-tailed test).

1) Natural ecosystem significantly affects the entire perception of harmonious tourism environment (0.32, p = 0.000). As the foundation of cultural

heritage and tourism activities, natural ecosystem plays the most important role in tourists' perception on tourism environment.

From the descriptive analysis shown in **Table 2**, tourists are relatively satisfied with the natural ecosystem. So, it is very necessary to sustain the harmony in natural ecosystem, as well as enhance the environment sanitation and monitoring.

2) Cultural ecosystem significantly affects the entire perception of harmonious tourism environment (0.24, p = 0.000). The surveys also show that tourists are satisfied with the conservation of cultural heritage, however the inheriting state of cultural traditions and the coordination of tourism facilities and environment are not very well in tourists' eyes. The path efficient between cultural ecosystem and the perception of tourism environment is relatively small, which also indicates that the cultural propagation function of cultural heritages have not been fully utilized and popular tourists could not pay enough attention on cultural heritages.

3) Social relation system significantly affects the entire perception of harmonious tourism environment (0.29, p = 0.000). It was found that the average score of social relation system is the lowest in three harmonious relation systems from the surveys, which indicates that tourists are not satisfied with the relationships among different stakeholders in "8 mountains". Because of the significance of the harmonious condition of stakeholders, it is very necessary to protect the legal rights and interests of local residents, alleviate the relation between residents and developers, hosts and guests, make tourism service more humanized, improve tourism merchant market and social security, as well as control the density of tourists.

4) Entire perception of harmonious tourism environment significantly and directly affects the tourists' loyalty (1.15, p = 0.000). Because of the very large coefficient (more than 1), in order to increases the tourists' satisfaction and loyalty degree, vigorously constructing the destination's harmonious tourism environment seems very fatal and crucial. We also can find that the tourists' recommendation intention is higher than revisit intentionso it is necessary to diversify the contemporary tourism products' structure, deepen the tourism products' cultural connotation, and make the tourists not visit only one time.

5. CONCLUSION

In all of the world heritages in China, the 8 mountaintype world cultural heritages are elites, because they represent the Chinese cultural spirits and nation's identity, as well as are the most famous tourist destinations, so the construction of harmonious tourism environment in "8 mountains" is very essential. In view of the very high positive coefficient of the entire perception of harmonious tourism environment to tourists' loyalty, we can conclude that harmonious tourism environment construction plays an essential role in enhancing the destination's competitiveness. Tourists now are satisfied with the natural ecosystem, which also most greatly affects the entire perception of harmonious tourism environment, thus it is necessary to maintain the harmony in natural ecosystem. Tourists' satisfaction on cultural ecosystem is relatively high, but the coefficient of cultural ecosystem to entire perception of harmonious tourism environment is relatively small, which indicates the defects exist in cultural tourism products and the low attention from tourists. Cultural heritages have very broad and deep cultural connotation, so it seems crucial for tourism developers to take more actions to diversify cultural tourism products and activities, broaden and deepen their cultural connotation, improve tourism demonstration and interpretation, as well as enhance cultural heritage education to tourists. The lowest satisfaction occurs in social relation system, shows the great flaw in the relationships among stakeholders in tourism developing process. Because research also shows the significant effect of social relation system to the perception of tourism environment, so it is very necessary and important to reform the management constitution, especially the interest distribution mechanism, protect local residents' legal rights and profits and enhance their participation intention in tourism development, eliminate the conflicts among stakeholders, finally construct a harmonious mechanism among residents, tourists and developers. There are still some drawbacks in this research, for instance, the samples inclined to select more young people, the quantity of questionnaires are not enough to make a comparative analysis among "8 mountains" etc, and some improvements are needed in future study.

6. ACKNOWLEDGEMENTS

This research was supported by Fund Project of National Social Science in China (Grant No. 07BJY136).

REFERENCES

1.   P. Murphy, M. P. Pritchard and B. Smith, "The Destination Product and Its Impact on Traveler Perceptions," Tourism Management, Vol. 21, No. 1, 2000, pp. 43-52.

2.   M. G. Gallarza and I. G. Saura, "Value Dimensions, Perceived Value, Satisfaction and Loyalty: An Investigation of University Students' Travel Behavior," Tourism Management, Vol. 27, No. 3, 2006, pp. 437-452.

3.   C. Lee, Y. Yoon and S. Lee, "Investigating the Relationships among Perceived Value, Satisfaction, and Recommendations: The Case of the Korean DMZ," Tourism Management, Vol. 28, No. 1, 2007, pp. 204-214.

4.   C. Y. Shi, J. Zhang and H. M. You, "Structural Equation Model for tourism Destination Competitiveness from Tourists' Perception Perspectives," Geographical Research, Vol. 27, No. 3, 2008, pp. 703-714.

5.   Y. L. Huang and Z. F. Huang, "A Study on the Structural Equation Model and Its Application to Tourist Perception for Agri-Tourism

Destinations: Taking Southwest Minority Areas as an Example," Geographical Research, Vol. 27, No. 6, 2008, pp. 1455-1465.

6.   Y. Yoon and M. Uysal, "An Examination of the Effects of Motivation and Satisfaction on Destination Loyalty: A Structural Model," Tourism Management, Vol. 26, No. 1, 2005, pp. 45-56.

7.   W. Zhang, J. W. Wang and H. Zhang, "Study on the Construction of Harmonious Tourism Environment for Mountain-Type World Cultural Heritages of China," Wuhan University Journal (Philosophy & Social Sciences), No. 2, 2008, pp. 278-282.

8.   M. Opperman, "Tourism Destination Loyalty," Journal of Travel Research, Vol. 39, No. 1, 2000, pp. 78-84.

9.   J. M. Fang, "An Empirical Study on the Incentive Influencing the Response Rate of the Web-Based Survey," Management Review, Vol. 18, No. 2, 2006, pp. 13-17.

10.   G. H. Chen and Y. S. Huang, "Influencing Factors on Tourists' Revisit Decision-Making: A Web-Based Empirical Study," Tourism Tribune, Vol. 23, No. 1, 2008, pp. 69-74.

CHAPTER 3

The Role and Importance of Cultural Tourism in Modern Tourism Industry

János Csapó

University of Pécs, Institute of Geography Hungary

1. INTRODUCTION

The main aim of this chapter is to thoroughly present the role and positions of cultural tourism, as one of modern tourism industry's most dynamically developing branch, in today's global tourism market both from the theoretical and the practical point of view.

With the definition of cultural tourism, we try to point at the complex problems of the term as it is proved to be a controversial issue in tourism, since there is no adequate definition existing. In the absence of a uniformly accepted definition, cultural tourism can be characterised both from the perspective of supply and demand and also from the point of view of theoretical and practical approach.

We can state that cultural tourism is a very complex segment of the 'tourism industry,' its supply is diverse and versatile. The future positions of the discipline will probably be strengthened both directly and indirectly as with the change of the recreational needs the aim to get acquainted with the cultural values is strongly increasing. Mass tourism though will of course never loose its positions, but tourists taking part in the supply of the 4S will become visitors with more diversified needs concerning cultural interest.

So apart from the theoretical discussion, the chapter aims to provide an insight into the tourism segments and attraction structure of cultural tourism as well.

1.1 The problems and definition of the term 'culture' and 'cultural tourism' 2.1 Defining the term 'culture'

To define cultural tourism first of all we have to determine the meaning of the term culture. In this chapter we do not intend to investigate this very complex concept from different aspects and approaches or with a very detailed analysis but we wish to provide an insight and a starting point since we feel that the determination of the context provides us the basics for the researches on cultural tourism.

So in this approach first of all we intend to highlight one of the first scholars who dealt with the identification of culture by providing a classic approach which is widely accepted in the scope of social sciences researchers. According to TYLOR (1871) culture is „that complex whole which includes knowledge, belief, art, morals, law, custom, and any other capabilities and habits acquired by man as a member of society" (Tylor, 1871.) This definition seems to be a favourable approach to our investigations as well since the determination can be used in a wide content opening the possibilities to the possible connection with other disciplines, and at the same time the definition is exact and concrete.

When analysing the meaning of culture we also would like to provide the approach and definition of the Webster's New Encyclopaedic Dictionary which states that culture is "the characteristic features of a civilisation including its beliefs, its artistic and material products, and its social institutions." (Webster's New Encyclopaedic Dictionary, p. 244).

On the other hand we also wish to explain that there is a strong and maybe ever lasting debate on the definition of this very complex term. Anthropology originally stated that culture and cultures are "unique bounded entities with limits and specific characteristics. Cultures were static, in that they could be captured by anthropological analyses. Their customs, habits, mores, relationships, uniquenesses could all be detailed, and in doing so, the ways in which each culture was separate from all others could be seen." (http://www.sccs.swarthmore.edu/users/00/ ckenned1/culture.html)

On the other hand recent trends of the research on culture show that culture is not a bounded, unchanging entity. Cultures are not separated from each other providing a chance to continuously interact and contact with each other. Of course this trend would also strongly determine the formation and development of cultural tourism as well.

From the more recent perspective we intend to highlight the definition of HOFSTEDE (1997) who states that: "Culture refers to the cumulative deposit of knowledge, experience, beliefs, values, attitudes, meanings, hierarchies, religion, notions of time, roles, spatial relations, concepts of the universe, and material objects and possessions acquired by a group of people in the course of generations through individual and group striving" (Hofstede, 1997).

According to HOFSTEDE (1997) the core of a culture is formed by the values (Figure 1.) which in terms of tourism will be the basics for the attraction of a given destinations well. The different levels of culture will be the rituals, the heroes and the symbols of the given culture which again would serve as a basis for tourism purpose travels.

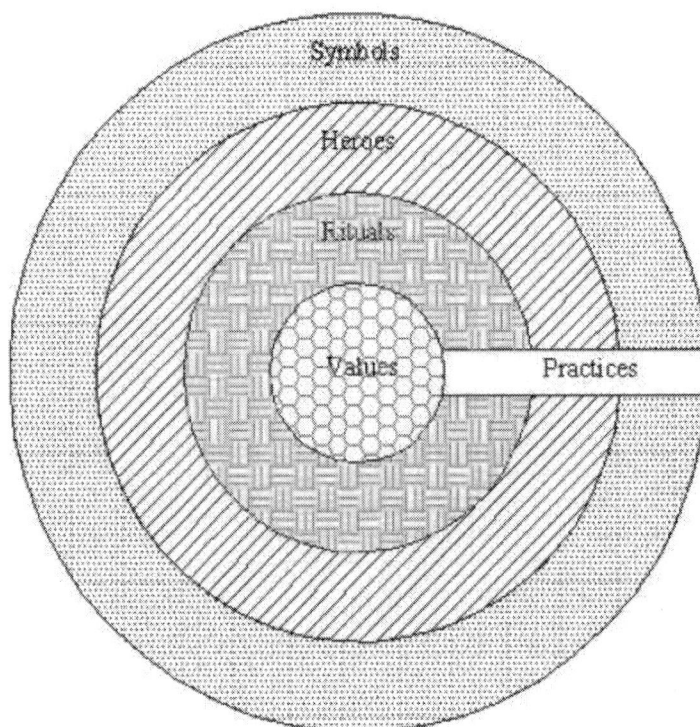

Figure 1. Manifestation of Culture at Different Levels of Depth (HOFSTEDE 1997)http://www.tamu.edu/faculty/choudhury/culture.html

We also agree with the definition of the Roshan Cultural Heritage Institute according to which "Culture refers to the following Ways of Life, including but not limited to:

- Language: the oldest human institution and the most sophisticated medium of expression.

- Arts & Sciences: the most advanced and refined forms of human expression.

- Thought: the ways in which people perceive, interpret, and understand the world around them.

- Spirituality: the value system transmitted through generations for the inner well-being of human beings, expressed through language and actions.

- Social activity: the shared pursuits within a cultural community, demonstrated in a variety of festivities and life-celebrating events.

- Interaction: the social aspects of human contact, including the give-and-take of socialization, negotiation, protocol, and conventions". (http://www.roshaninstitute.org/474552)

Based on the above mentioned we can state that culture is part of the lifestyle which a multitude of people are sharing. The similarities in spoken and written language, behaviour, lifestyle, customs, heritage, ideology and even technology connect the individuals to groups of people in a certain culture. So now if we take into consideration cultural tourism these groups will constitute on the demand side on the one hand those tourists who are possessing cultural motivation during their travel and on the other hand from the supply side the destination which is disposing those attraction which are capable to desire the attraction of a culturally motivated tourists or visitor. So based on the upper mentioned we could also state that the altering explanations of cultural tourism could also be derived from the altering meanings and interpretations of the term culture.

1.2 Defining the term 'cultural tourism'

The concept of cultural tourism again is very complex and so there is a long debate among scholars about its definition and conceptualisation (Michalkó, 2004; Richards 2005; Shackleford, 2001) due to which we find numerous definitions for this term. So as one of the most important recent papers on cultural tourism – more exactly cultural city tourism – mentions, "there are a great number of definitions of cultural tourism in use, resulting in different definitions being used in research studies related to cultural tourism and in the field of cultural tourism." (City Tourism and Culture – The European Experience, 2005)

We can clearly see that this approach and the practice itself proves that the discourse on cultural tourism is extremely difficult which could result in false understanding of the term and also – from the point of view of the practical approach – we could highlight that for instance statistical background and research of this discipline seems to be more and more difficult due to the mentioned phenomena. As McKercher and Du Cros (2002) responds to the question: "What is cultural tourism? This seemingly simple question is actually very difficult to answer because there are almost as many definitions of cultural tourism as there are cultural tourists." (McKercher & Du Cros 2002)

When starting with the definitions first we would like to mention the Dictionary of Travel, Tourism and Hospitality Terms published in 1996 according to which "Cultural tourism: General term referring to leisure travel motivated by one or more aspects of the culture of a particular area." ('Dictionary of Travel, Tourism and Hospitality Terms', 1996).

One of the most diverse and specific definitions from the 1990s is provided by ICOMOS (International Scientific Committee on Cultural Tourism): "Cultural tourism can be defined as that activity which enables people to experience the different ways of life of other people, thereby gaining at first hand an understanding of their customs, traditions, the physical environment, the intellectual ideas and those places of architectural, historic, archaeological or other cultural significance which remain from earlier times. Cultural tourism differs from recreational tourism in that it seeks to gain an understanding or appreciation of the nature of the place being visited." (ICOMOS Charter for Cultural Tourism, Draft April 1997). We strongly accept and favour this definition which on the one hand seems to be a bit too long, but mentions and highlights not just the man made attractions connected to cultural tourism, but the surrounding physical environment as well providing a wider spatial scope to this form of tourism.

It is also interesting to mention that the definition has been improved through the years of the committee's practice since their 1976 definition was somewhat simpler and not that precise than the previously mentioned one: "Cultural tourism is that form of tourism whose object is, among other aims, the discovery of monuments and sites. It exerts on these last a very positive effect insofar as it contributes - to satisfy its own ends - to their maintenance and protection. This form of tourism justifies in fact the efforts which said maintenance and protection demand of the human community because of the socio-cultural and economic benefits which they bestow on all the populations concerned." (1976 ICOMOS Charter on Cultural Tourism)

There are other definitions from this era which focus on one of the most important effects of tourism on the tourists, namely the experiences. One of these definitions were set up by Australian Office of National Tourism: "Cultural tourism is tourism that focuses on the culture of a destination - the lifestyle, heritage, arts, industries and leisure pursuits of the local population." (Office of National Tourism 'Fact Sheet No 10 Cultural Tourism', 1997). The earlier mentioned charter of the ICOMOS describes cultural tourism as: "Cultural tourism may be defined as that movement which involves people in the exploration or the experience of the diverse ways of life of other people, reflecting all the social customs, religious traditions, or intellectual ideas of their cultural heritage." (ICOMOS Charter for Cultural Tourism, Draft April 1997).

We provide two more definitions focusing on experience during the trip:

"Cultural tourism is an entertainment and educational experience that combines the arts with natural and social heritage and history." (Cultural Tourism Industry Group, http://www.culturaltourismvictoria.com.au/).

So we see that some of the definitions try to focus on the attraction side of this system, some on the geographical space and some on the experiences but fortunately almost all of them focus on and highlight the role of the local population as well.

Even there are some country or space specific definitions for cultural tourism such as in Australia: "Cultural tourism is defined by attendance by inbound visitors at one or more of the following cultural attractions during their visit to Australia: festivals or fairs (music, dance, comedy, visual arts, multi-arts and heritage); performing arts or concerts (theatre, opera, ballet and classical and contemporary music); museums or art galleries; historic or heritage buildings, sites or monuments; art or craft workshops or studios; and Aboriginal sites and cultural displays." (Bureau of Tourism Research, 'Cultural Tourism in Australia', 1998, p.7).

One of the most important professional initiatives of cultural tourism is provided by the ATLAS Cultural Tourism Research Project which was aiming to establish a transnational database which could provide comparative data on cultural tourism trends across Europe (Bonink et al. 1994). Due to its more than 15 years of activity the ATLAS Cultural Tourism Research Programme has monitored one of the most rapidly growing areas of global tourism demand through visitor survey and studies of cultural tourism policies and suppliers (http://www.tram-research.com/atlas/presentation.htm). The ATLAS program provides us two new definitions from a conceptual and a technical perspective:

CONCEPTUAL DEFINITION

"The movement of persons to cultural attractions away from their normal place of residence, with the intention to gather new information and experiences to satisfy their cultural needs".

TECHNICAL DEFINITION

"All movements of persons to specific cultural attractions, such as heritage sites, artistic and cultural manifestations, arts and drama outside their normal place of residence". (ATLAS, 2009)

When taking into consideration the definition of the term cultural tourism of course we highlight the approach of the UNWTO. The United Nations World Travel Organisation provides us two perspectives of the definition of cultural tourism, namely a broad and a narrow approach:

- *"All movements of persons might be included in the definition because they satisfy the human need for diversity, tending to raise the cultural level of the individual and giving rise to new knowledge, experience and encounters. (broad definition).*

- *Movements of persons for essentially cultural motivations such as study tours, performing arts and cultural tours, travel to festivals and other cultural events, visits to sites and monuments. (narrow definition)."* (UNWTO)

The broad approach can hardly be handled from the point of view of product development and product management aspects since in this respect almost all the recreational travels could be ranged to the scope of cultural tourism as due to the new experiences the tourist will realize new observations and knowledge (Michalkó & Rátz 2011).

If we take into consideration the narrow sense of the UNWTO's definition the programs, events and sightseeings of the so called high or elite culture provides the basic attraction for cultural tourism. In this respect monuments and heritage sites, festival tourism, exhibitions and museums, visiting theatres and concerts and pilgrimage or study tours are the basic products of cultural tourism.

According to MICHALKÓ and RÁTZ – in accordance with our perceptions as well – one has to take into consideration the popular culture also when investigating cultural tourism. In this respect we can highlight such tourism products as rock or pop music festivals, or "movie" tourism (visiting places where famous films were shot) as well (Michalkó & Rátz 2011).

Based on the above mentioned the definition of the two tourism researchers on cultural tourism is the following: *"Cultural tourism is such a tourism product in which the motivation of the tourist (providing the supply side) is getting acquainted with new cultures, participate in cultural events and visiting cultural attractions and the demand side's core element is the peculiar, unique culture of the visited destination"*. (Translated by the authors from Hungarian) (Michalkó & Rátz 2011).

The 2005 report of the European Travel Commission on City Tourism and Culture distinguishes between an inner and outer circle of cultural tourism:

- *"I. The inner circle represents the primary elements of cultural tourism which can be divided into two parts, namely heritage tourism (cultural heritage related to artefacts of the past) and arts tourism (related to contemporary cultural production such as the performing and visual arts, contemporary architecture, literature, etc.).*

- *II. The outer circle represents the secondary elements of cultural tourism which can be divided into two elements, namely lifestyle (elements such as beliefs, cuisine, traditions, folklore, etc.) and the creative industries (fashion design, web and graphic design, film, media and entertainment, etc.)."* (City Tourism and Culture – The European Experience, 2005)

Here we also would like to mention and introduce the widely accepted definition of Stebbins (1996) who states that "Cultural tourism is a genre of special interest tourism based on the search for and participation in new and deep cultural experiences, whether aesthetic, intellectual, emotional, or psychological." (Stebbins, 1996)

Without the aim of listing all the definitions on cultural tourism, we would like to emphasize that according to our point of view the scope of cultural tourism covers those tourism segments that could not be classified to the elements of mass and passive tourism. The classic attractions of cultural tourism can be classified into three groups:

- Built and material values (buildings, material values of different art forms),

- The cultural values connected to everyday life (free time, leisure, lifestyle, habits, gastronomy,

- Events and festivals (Aubert & Csapó 2002).

According to our latest knowledge and as an edification from the above mentioned definitions we should presume that the definitions of culture and tourism reflects together the meaning of cultural tourism. In this case this part or area of tourism is a collecting concept which is multiple and diversified from the point of view of several tourism products – with cultural attraction – which are determined in the next chapter.

1.3 The typology of the cultural tourist

When dealing with the very complex phenomenon of cultural tourism it is also necessary to determine who is a cultural tourist. Based on the above mentioned chapters, according to our point of view, such a tourist takes part in cultural tourism who is not travelling away from home to reproduce the needs and necessities of the home environment in more advantageous and desirable circumstances in a remote land or country but he or she is disposed with the adequate (cultural) motivation getting to know the different and remote (local) culture's social and landscape values. We believe that apart from the – more traditionally 'used' social cultural values – we should also highlight the role of the natural environment concerning cultural tourism.

When we defined who is a cultural tourist the next step in our research would be the typology of those taking part in cultural tourism. This typology seems to be almost as complex as the previous definitions. In our work we accept and favour the typology of McKercher and Du Gros who differentiated five types of cultural tourists based on the importance of culture in their decision to travel and also the depth of their experience (*Figure 2.*).

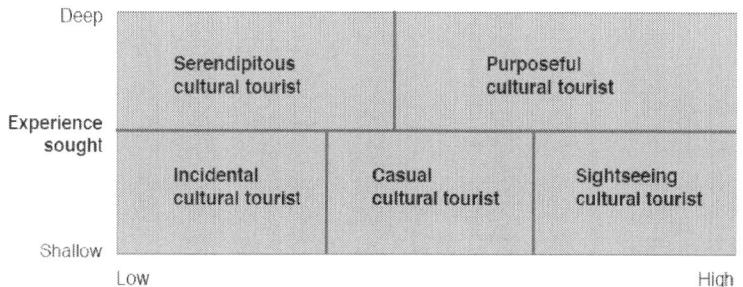

Source: City Tourism & Culture - The European Experience, p. 4.

Figure 2. The typology of cultural tourist by McKercher and Du Cros

As we have already seen, tourists can be totally, partially or only incidentally be involved in cultural tourism or in culturally motivated activities. So it is natural that due to this phenomenon we believe that statistically it is very hard to register tourists belonging to whichever category of tourism activities. We can distinguish between specific and incidental cultural tourists but we also have to stress that the boundary between each categories is very hard to be determined as well.

Table 1. Types of cultural tourists by McKercher and Du Cros

Type of cultural tourist	Short characterisation
The purposeful cultural tourist	Cultural tourism is the primary motivation for visiting a destination and the tourist has a very deep and elaborate cultural experience
The sightseeing cultural tourist	Cultural tourism is a primary reason for visiting a destination, but the experience is less deep and elaborated
The serendipitous cultural tourist	A tourist who does not travel for cultural reasons, but who, after participating, ends up having a deep cultural tourism experience
The casual cultural tourist	Cultural tourism is a weak motive for travel and the resulting experience is shallow
The incidental cultural tourist	This tourist does not travel for cultural reasons, but nonetheless participates in some activities and has shallow experiences

Source: With minor alterations by the author, based on City Tourism & Culture – The European Experience, p. 4. own editing

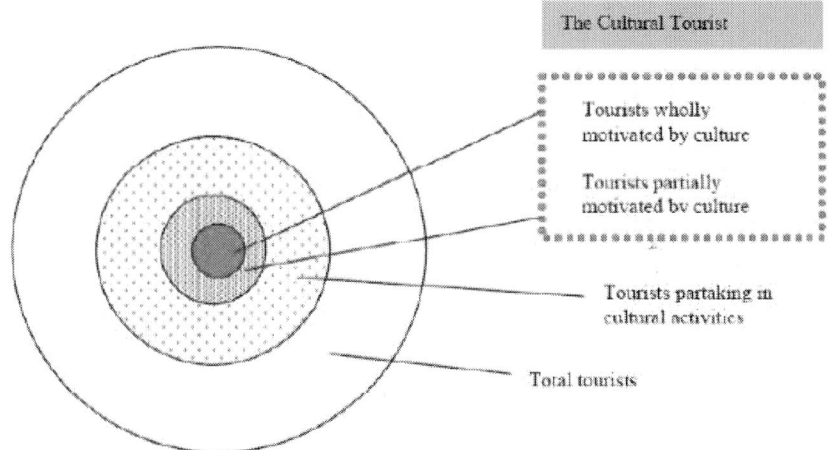

Figure 3. The place of cultural tourists in the complete tourist flow

It is also to be stressed that taking into consideration the number of tourists involved in cultural tourism, so in other words the quantitative aspects, the most of the tourists will be registered to the 'incidental' category and the least amount of people will travel to an attraction or destination with a 100% of cultural motivation.

1.4 Types of cultural tourism

In this chapter we intend to introduce and highlight the most important types or in other words elements of cultural tourism from a thematic perspective grouped by the principles of the preferred activity. According to our standardisation we classify cultural tourism in the following way:

Table 2. Classification of major cultural tourism forms

Types of cultural tourism	Tourism products, activities
Heritage tourism	• Natural and cultural heritage (very much connected to nature-based or ecotourism); • Material - built heritage, - architectural sites, - world heritage sites, - national and historical memorials • Non material - literature, - arts, - folklore • Cultural heritage sites - museums, collections, - libraries, - theatres, - event locations, - memories connected to historical persons
Cultural thematic routes	• wide range of themes and types: - spiritual, - industrial, - artistic, - gastronomic, - architectural, - linguistic, - vernacular, - minority
Cultural city tourism, cultural tours	• "classic" city tourism, sightseeing • Cultural Capitals of Europe • "Cities as creative spaces for cultural tourism"
Traditions, ethnic tourism	• Local cultures' traditions • Ethnic diversity
Event and festival tourism	• Cultural festivals and events - Music festivals and events (classic and light or pop music) - Fine arts festivals and events

Types of cultural tourism	Tourism products, activities
Religious tourism, pilgrimage routes	<ul><li>Visiting religious sites and locations with religious motivation</li><li>Visiting religious sites and locations without religious motivation (desired by the architectural and cultural importance of the sight)</li><li>Pilgrimage routes</li></ul>
Creative culture, creative tourism	<ul><li>traditional cultural and artistic activities</li></ul> - performing arts, - visual arts, - cultural heritage and literature <ul><li>as well as cultural industries</li></ul>- printed works, - multimedia, - the press, - cinema, - audiovisual and phonographic productions, - craft, - design and cultural tourism

Ed. Csapó. J. 2011

The major (directly) connected tourism products for cultural tourism are rural tourism (traditions, lifestyle, local gastronomy), wine tourism (grape and viticulture), conference tourism and eco-tourism (local culture, lifestyle).

In the following parts of the chapter we try to focus on and introduce the major tourism products that can be related to and so characterising cultural tourism.

Heritage tourism

Heritage tourism and its different forms as mentioned in the table above mean nowadays one of the most important forms of cultural tourism. *"Thanks to a global, integrated approach in which nature meets culture, the past meets the present, the monumental and movable heritage meets the intangible, the protection of cultural heritage, as an expression of living culture, contributes to the development of societies and the building of peace. By virtue of its multifarious origins and the various influences that have shaped it throughout history, cultural heritage takes different tangible and intangible forms, all of which are invaluable for cultural diversity as the wellspring of wealth and creativity."* (http://www.unesco.org/en/culturaldiversity/heritage/)

Based on the above mentioned – in accordance with the definition of the National Trust for Historic Preservation's Heritage Tourism Program (http://culturalheritagetourism. org/documents/2011CHTFactSheet6-11.pdf) – we would state that heritage tourism is an important part of cultural tourism based on experiencing the places and activities that authentically represent historic, cultural and natural resources of a given area of region.

Taking into consideration the classification of cultural tourism, The Unites Nations Educational, Scientific and Cultural Organisation (UNESCO) differentiates different types of heritage such as monumental, movable, intangible and world heritage.

If we take into consideration the forms of heritage and heritage tourism we can differentiate between material (built heritage, architectural sites, world heritage sites, national and historical memorials) and non-material heritage (literature, arts, folklore) and cultural heritage sites such as museums, collections, libraries, theatres, event locations and memories connected to historical persons.

We also agree with the identification and classification of Timothy and Boyd (2003) stating that "heritage can be classified as tangible immovable resources (e.g. buildings, rivers, natural areas); tangible movable resources (e.g. objects in museums, documents in archives); or intangibles such as values, customs, ceremonies, lifestyles, and including experiences such as festivals, arts and cultural events". (Timothy & Boyd, 2003)

Heritage tourism is quite a new phenomena on the one hand concerning cultural tourism but on the other hand its routes can be traced back to the ancient times of human history. Due to the modern trends of tourism its demand has been rapidly growing from the 1990s but especially in the 21st century.

Of course in the focus of heritage tourism it is heritage itself which mean such a cultural value from the past which is worth to be maintained for the new generations. Within heritage, we can differentiate between natural and cultural heritage as well. So when we would like to define heritage tourism it is essential to highlight that it is such a form of tourism that is based on heritage in which heritage is one the one hand the central element of the tourism product and on the other hand it provides the major motivation for the tourist. (Swarbrooke, 1994).

The recent trends of the extraordinary growth of heritage tourism development are due to several phenomena experienced in social life and the trends of the tourism industry:

- The media participates more and more acutely in introducing the heritage sites;

- By the increase of the education level of the population an increasing need has been emerged to travels with cultural (heritage) purposes;

- Heritages became a product consumable for the tourist due to the intermediary role of the tourism industry;

- The personal and social value and support of heritage and heritage tourism has grown from the second half of the 20th century. (Berki, 2004)

By the end of the 20th century and the beginning of the 21st century new trends have emerged in heritage tourism as well. There were significant changes on the fields of heritage attractions, the need for complex tourism products also has been grown on the demand side and so the traditional cultural attractions (such as museums) had to revalue their original role. (Richards, 2001)

The characteristic segments for the modern heritage tourism are the following:

- Tourists are represented mainly with a higher educational background;

- The specific spendings of these tourists are higher than average;

- Tourists are rather coming from the urbanised areas and from the more developed "western world".

- Their majority is in their middle ages without children;

- According to the length of stay we can state that in the case of heritage tourism the time for the travel is shorter while the frequency of the travels is higher. (Berki, 2004)

World heritage

World heritage sites were created by the UNESCO's Convention Concerning The Protection Of The World Cultural And Natural Heritage adopted by the General Conference at its seventeenth session in Paris, 16 November 1972.

Figure 4. The logo of the UNESCO's World Heritage Sites

One of the major driving force for this decision was that the convention noted *"that the cultural heritage and the natural heritage are increasingly threatened with destruction not only by the traditional causes of decay, but also by changing social and economic conditions which aggravate the situation with even more formidable phenomena of damage or destruction."* http://whc.unesco.org/en/conventiontext/

Since this decision in 1972 by today the World Heritage List includes 936 properties both from cultural and natural heritage which the World Heritage Committee considers as having outstanding universal value. http://whc.unesco.org/en/list/ By 2011, this number is classified into 725 cultural, 183 natural and 28 mixed properties in 153 States Parties.

Table 3. UNESCO's definitions for cultural heritage and natural heritage

Article 1 For the purposes of this Convention, the following shall be considered as "cultural heritage":	
monuments:	architectural works, works of monumental sculpture and painting, elements or structures of an archaeological nature, inscriptions, cave dwellings and combinations of features, which are of outstanding universal value from the point of view of history, art or science;
groups of buildings:	groups of separate or connected buildings which, because of their architecture, their homogeneity or their place in the landscape, are of outstanding universal value from the point of view of history, art or science;
sites:	works of man or the combined works of nature and man, and areas including archaeological sites which are of outstanding universal value from the historical, aesthetic, ethnological or anthropological point of view.

Article 2 For the purposes of this Convention, the following shall be considered as "natural heritage":	
natural features	consisting of physical and biological formations or groups of such formations, which are of outstanding universal value from the aesthetic or scientific point of view;
geological and physiographical formations and precisely delineated areas	which constitute the habitat of threatened species of animals and plants of outstanding universal value from the point of view of science or conservation;
natural sites	or precisely delineated natural areas of outstanding universal value from the point of view of science, conservation or natural beauty.

Source: based on http://whc.unesco.org/en/conventiontext/ edited by Csapó, 2011.

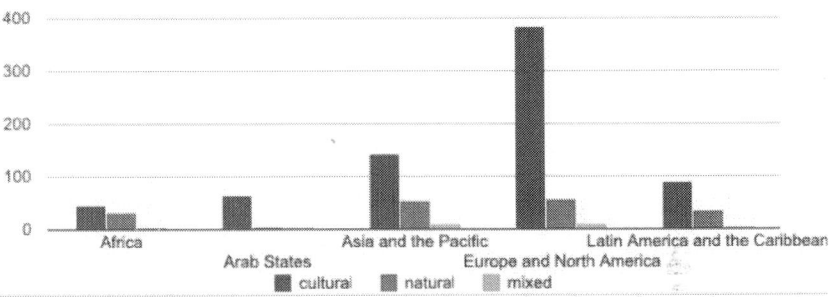

Source: http://whc.unesco.org/pg.cfm?cid=31&l=en&action=stat&&&mode=table

Figure 5. Number of World Heritage properties by region

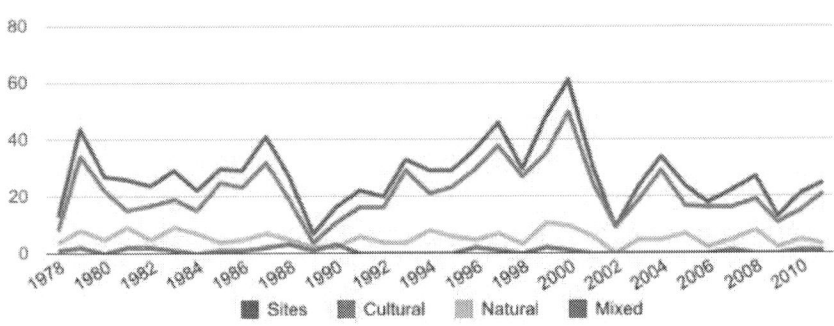

Source: http://whc.unesco.org/pg.cfm?cid=31&l=en&action=stat&&&mode=table

Figure 6. Number of World Heritage properties inscribed each Year

The role and importance of thematic routes in cultural tourism

The concept and definition of the thematic routes

The direction of the tourism supply development moved towards thematic supplies in the 1980s first in Western-Europe, in the United States of America and in Australia, then in the second part of the decade in East Central Europe and other regions as well.

The thematic supply development means such a planning and realisation that is adjusted to attraction features and uniqueness including all the services that the tourist presses into service. The accentuation of the given region's featuring attractions comes into prominence. The forming supplies and their originated travel products chose such a highlighted feature of the rural regions which are able to represent independently the attraction of the given area. In many cases however in the core of thematic attraction development there is an artificial attraction (Aubert & Csapó 2002; Berki & Csapó, 2008).

The forming of the thematic routes can be reckoned among the methods of the thematic supply development, so the foundation of the thematic parks, or the destination supply development of the close sense.

The thematic routes – according to Puczkó & Rátz (2000; 2007) – are such tourism products which row up natural or man made attractions accessible by different transport forms around a chosen topic or theme. When developing thematic routes, the more increased application of the given attractions is a general aim due to which this supply will be more strongly taking part in the tourism of the given region and area.

The participants of the co-operations in the initial period were the operators and proprietors of the attractions – so in a number of cases local governments or organisations owned by the local governments – to which later on the enterprises of the competition sector joined as well. Besides this, to the lower level of hierarchy the co-operation with only marketing functions is characteristic. In this case the aim of the given characters is the increasing of the efficiency of the advertisements besides the reduction of the specific advertisement expenses. On the higher organization level of co-operations with extended activities a standardisation process is experienced with creating a common image. The appearance of such a supply supposes the creation of travel packages as well due to the connecting attractions. (Berki & Csapó, 2008)

On the successful operation of the thematic routes we find numerous successful methods and examples in an outside Europe. Forming an international co-operation may have a number of advantages but challenges as well to the participants. The creation of the route is seemingly an easy task so the attractions have to be selected and developed adequate to the main theme, and applying management methods as well. It can be referred to the positive effects that considering costs these supplies are created with a small range of investment, they be diverse both spatially and timely, can contribute to the unutilised tourism resources and can captivate a new demand group for the cultural and heritage tourism.

We keep count on the benefits side of thematic route creation that

- They can realised with a relatively small investment,
- Are able to diverse the tourism demand both timely and spatially,
- They are able to utilize unexploited resources,
- A new demand group can be captivated to the given attraction.

Apart from the above mentioned we may interpret as an additional positive economic effect,

- the motivation of the enterprises among local residents,
- support of investments, and development concerning buildings, infrastructure and human resources,
- the settling down of related services, which can also be used by the local residents,
- the effect of the income increase due to the increasing tourism flow,
- and as a result of the above mentioned workplace creation. (Berki & Csapó, 2008)

At attractions functioning in the long haul we can find concrete results so the spatial development effect of tourism can be shown as well. Apart from the economic effects the social effects could be of great importance as well, so as the promotion of the connection system between communities and culture.

Standardised thematic routes on high level of hierarchy with common image presenting international co-operation

Here in this section we would like to highlight both cultural and natural attraction-based co-operations that are characteristic and greatly important in Europe's thematic route structure.

Cultural routes in Europe

By far the most important segment of thematic routes is the cultural routes. The first cultural routes were introduced by the Council of Europe in 1987. This cooperation of the Council started first with ten cultural routes marking the stages in Europe's development, realising that Europe's cultural routes cross over and link the local, the regional and the international level as well.

This co-operation for today is one of the most well known and well practicing in Europe. As a result of the experienced fast development in 1998 the Council of Europe wanted to set the project within a more formal co-operation framework by establishing certain Regulations and so created the Atlas of Cultural Routes. These regulations established a reformed network of cultural routes in Europe. (European Institute of Cultural Routes)

The following list of cultural routes is a collection of the most important routes in Europe with a high level of hierarchy and tourist image. (Berki & Csapó, 2008)

European Route of Industrial Heritage (IRIH)

Another remarkable – although future – co-operation of European thematic routes is the European Route of Industrial Heritage with an altogether of thirteen partners from Belgium, Germany, the Netherlands and the UK with jointly invest of €2.6 Million to develop together the first European network dedicated to industrial heritage.

Table 4. A List of Cultural Routes in Europe

Transnational Cultural Routes
The Via Francigena Federation of the Clunisian sites in Europe The Way of St. James Klösterreich

Romanesque Routes of Europe	
Romanesque road in Saxony-Anhalt (Germany) Romanesque itineraries in Thuringia (Germany) Empire and region - Saxony-Anhalt (Germany) Romanesque art and architecture in Bavaria (Germany) Romanesque churches in Cologne (Germany) Ways to the Romanesque in Lower Saxony (Germany) Romanesque in Osnabrück (Germany) Romanesque itineraries in South Burgundy (France) Romanesque art in Poitou-Charentes (France) Historical routes of the medieval abbeys in Normandy (France)	Romanesque road in Alsace (France) Romanesque art in the Provence (France) Romanesque art and architecture in France Romanesque in Spain Romanesque art in Soria (Spain) Romanesque art in Spain Fundación Uncastillo Romanesque route in the province of Asti (Italy) Romanesque route in Middle Poland The Romanesque album (Poland)

Further thematic itineraries
The Saint James Way through Saxony-Anhalt The Saint James Way through Germany, Switzerland, France and Spain Open churches in the middle of Germany The Via Imperialis Villas - Stately Homes and Castles in Carinthia (Austria) Città d´arte in Emilia-Romagna (Italy) Sentieri della Luce -Matilda Path (Italy) The Lands of Mathilde of Canossa (Italy) Castles of the duc of Parma and Piacenza (Italy) Pilgrimages in Carinthia (Austria) The Romans route The Route of the Castillian language

Source: http://www.culture-routes.lu

This five year project is led by a German tourist board, building on the findings of a previous project under IIC to demonstrate that industrial heritage can be a valuable resource. The highlighted geographical areas are the German Rühr, the North of England and the Saar-Loor-Lux area. The thematic routes will be developed at transnational level around former industries such as textile, mining or steel. By establishing the so called "Anchor Points" of these areas people will be attracted by 60 important industrial heritage sites with a well developed tourism infrastructure. (http://en.erih.net/)

This route system consists of thematic- and regional routes as well (*Table 5*).

Table 5. Thematic and Regional Routes of the IRIH project

Theme Routes	Regional Routes
Theme Route Textiles	
Theme Route Mining	Ruhrgebiet (Germany)
Theme Route Iron and Steel	Industrial Valleys (Germany)
Theme Route Manufacturing	Euregio Maas-Rhine (Germany)
Theme Route Energy	Saar-Lor-Lux (Germany)
Theme Route Transport and	Lusatia (Germany)
Communication	Northwest England (Great Britain)
Theme Route Water	Heart of England (Great Britain)
Theme Route Housing and Architecture	South Wales (Great Britain)
Theme Route Service and Leisure	The Industrious East (Great Britain)
Industry	HollandRoute (The Netherlands)
Theme Route Industrial Landscapes	

Source: http://en.erih.net/

Heritage Tour project

This international project on cultural thematic route development in rural areas is cofinanced by the European Union within the INTERREG IIIB CADSES Programme. The project's "main objective is the protection, thematic organisation and promotion of local cultural heritage in remote/rural/mountainous/border areas in forms of regional and transnational cultural routes. The project's long term objective is the preservation of local cultural heritage in European villages, and the economic development of rural areas of the EU by assessing and developing their local cultural values into a tourist attraction, providing a good basis for further development of rural tourism." (HeriTour http://project.heritour.com/)

However the programme is only launched in 2007 we consider it as an important representative for international and cross-border co-operation. The four thematic routes of this programme will be based on Church history, industrial traditions, natural values and folk traditions. (Berki & Csapó, 2008)

City tourism

European cities and cultural tourism

Another segment of cultural tourism is city tourism or more precisely city tourism with cultural purposes. Without the aim of profoundly introducing this form of tourism we would like to highlight that still a certain proportion of the cultural tourism arrivals are motivated by city tourism (Michalkó, 1999). According to one of the most important researches in this respect we intend to highlight the scientific results of the World Tourism Organization and European Travel Commission carried out in 2005 entitled City Tourism & Culture - The European Experience.

According to the supply side the research paper classified the European settlements into six groups or clusters with the following major characteristics:

- "Villages only offer cultural heritage (cluster 1) and no or very limited visual arts, performing arts or the creative industries;

- Towns offer cultural heritage (cluster 2) and the visual and/or performing arts (cluster 3), but no or very limited creative industries;

- Cities offer cultural heritage and the performing and/or visual arts (cluster 4) and possibly the creative industries (cluster 5);

- Metropolises offer cultural heritage and the performing and/or visual arts and the creative industries (cluster 6)." (City Tourism and Culture, p. 6.)

Product category \ Type of place	Village	Town	City	Metropolis
Heritage	Cluster 1	Cluster 2		
Heritage + The Arts		Cluster 3	Cluster 4	
Heritage + The Arts + Creative Industries			Cluster 5	Cluster 6

Source: City Tourism and Culture, p.5.

Figure 7. A framework to classify places and their cultural product

Table 6. Classification of some European cities according to the framework

Cluster 2	Cluster 3	Cluster 4	Cluster 5	Cluster 6
Ávila	Avignon			
Bamberg	Basel			
Bern	Bayreuth	Athens	Amsterdam	
Canterbury	Bologna	Antwerp	Barcelona	
Córdoba	Bratislava	Bucharest	Brussels	
Delft	Bruges	Edinburgh	Budapest	
Granada	Florence	Glasgow	Copenhagen	Berlin
Heidelberg	Gent	Hamburg	Dublin	Istanbul
Luxembourg	Krakow	Helsinki	Lisbon	London
Oxford	Ljubljana	Porto	Lyon	Madrid
Monaco	Oslo	Prague	Milan	Paris
Nicosia	Santiago de	Riga	Munich	Rome
Pisa	Compostela	Rotterdam	Naples	
Siena	Sofia	Salzburg	Stockholm	
Valleta	Tallin	Seville	Vienna	
Würzburg	Venice	Warsaw		
York	Vilnius			
	Zagreb			

Source: City Tourism and Culture, p. 6.

European Capital of Culture programme

Table 7. The European Capitals of Culture (1985-2014)

1985: Athens
1986: Florence
1987: Amsterdam
1988: Berlin
1989: Paris
1990: Glasgow
1991: Dublin
1992: Madrid
1993: Antwerp
1994: Lisbon
1995: Luxembourg
1996: Copenhagen
1997: Thessaloniki
1998: Stockholm
1999: Weimar
2000: Avignon, Bergen, Bologna, Brussels, Helsinki, Krakow, Reykjavik, Prague, Santiago de Compostela.
2001: Porto and Rotterdam
2002: Bruges and Salamanca
2003: Graz
2004: Genoa and Lille
2005: Cork
2006: Patras
2007: Luxembourg and Sibiu
2008: Liverpool and Stavanger
2009: Linz and Vilnius
2010: Essen for the Ruhr, Pécs, Istanbul
2011: Turku and Tallinn
2012: Guimarães and Maribor
2013: Marseille and Kosice
2014: Umeå and Riga

Source: http://ec.europa.eu/culture/our-programmes-and-actions/doc2485_en.htm

Throughout the years this initiative was not only fostering the 'cultural industry' of the cities but they were a driving force regenerate the cities, raise the international profile of the ECC's and enhance their image from the point fo view of their own inhabitants. They also gave new vitality to their cultural life, raised their international profile, boosted tourism and enhanced their image in the eyes of their own inhabitants. http://ec.europa.eu/culture/ our-programmes-and-actions/doc413_en.htm

Since 1985 there was a debate and research on what impacts the ECC title had on the development of the cities or on the change of their image etc. People and researchers were of course criticising the major events or thematics one city was preparing for the given year. Not considering and arguing with these approaches we have to state that so far this initiative became one of the most important ones from the European Union in order to achieve a certain development on the image change and cultural development of major cities in Europe.

Traditions, ethnic tourism

According to our point of view we differentiate two types of ethnic tourism. One of them is "root tourism" and the other – more widely in practice - is tourism with the purpose of getting to know other people's differing cultural background from an authentic approach. According to SANYAL (2009) ethnic tourism is "travel motivated by search for the first hand, authentic and sometimes intimate contact with people whose ethnic and /or cultural background is different from the tourists". http://anandasanyal.blogspot.com/2009/06/ethnic-tourism-istravel-motivated-by.html So visitors with ethnic cultural motivations travel to another destination in order to be acquainted with a different culture. One of the major motivations in this travel is of course curiosity and also respect to other ethnic groups.

Within ethnic tourism we can differentiate between anthropological and tribal tourism as well, but village tourism (where the living conditions and again the different cultural approaches to every day life can be studied) should be classified here as well. One of the most important advantages of ethnic tourism is that this form of travel can be studied and experienced in almost every part of the world providing a great opportunity for the conservation of culture and heritage and also as tourism is the 'industry of peace', people's tolerance and cultural understanding could lead to a more peaceful approach to modern life and the negative impacts of globalisation as well.

A special form of ethnic tourism is root tourism where the driving force for travel is getting to know the culture of someone's (long-ago) homeland, either originated from the given area or being one of the offspring of someone. Such an example is perfectly presented in Ireland where the quest for the ancient homeland produced a complete tourism industry in the Republic of Ireland supported by the huge masses of the Irish diaspora all over the world but especially from the United States of America. (Trócsányi & Csapó, 2002) Genealogy research is one of the most sensational form of this root-researches. Such examples can be studied of course everywhere in the world where history brought some changes in the country borders (e.g. Hungary) or in certain periods of time masses of people were migrating away from the home country (European countries to the USA in the 1920s, 1930s).

We also would like to stress that one of the most important aims and objectives of this form of tourism is to get to know other's culture without disturbing and negatively effecting the local population because there is a threat in the development of this form of tourism that it leads to mass tourism with all its negative effects on the local culture and population.

Religious tourism, pilgrimage routes

Religious tourism and pilgrimage routes are the most ancient forms of tourism. If we take into consideration religion as a motivation we have to state that under religious tourism we understand the following activities:

- Visiting religious sites and monuments (churches, clusters, exhibition places)

- Taking part in religious events (holy days, religious cultural and music programmes, visiting religious persons)

- Pilgrimage

- Spiritual training (youth camps, missions etc.) (Nyíri, 2004)

- So we can differentiate between different groups of travellers with religious motivation such as:

- Organised groups visiting sacred places as a tourism destinations (either with religious motivation or with a motivation desired by the architectural and cultural importance of the sight)

- Individually organised visitors with their own programme organisation

- Such cultural tourists who have unique interests

- Pilgrims who are attending in an organised way for spiritual training

- Pilgrims who are attending individually for their spiritual training.

Recent researches show that this segment of cultural tourism has produced a tremendous growth especially from the 1990s. The number of visitors taking part in religious tourism and their tourism spendings totalled an estimated 18 billion USD with over 300 million travellers worldwide. (http://www.travelindustrydeals.com/news/5041) It also raises the question of carrying capacity since many of the religious sites are simply unable to bear the amount of people trying to visit the places either connected to any of the world's most important religions. So the great world religions have a high base for the massive religious or pilgrimage tours since millions of people are attracted to visit their sacred places or events. (Csapó & Matesz, 2007)

Event and festival tourism

Cultural events and festivals again play an important role in the formation and strengthening of cultural tourism in today's tourism industry. These programmes *"offer the tourist additional reasons to visit a place over and above the regular cultural product offered. Often because events are one-off and take place in a limited timeframe and because festivals offer a concentrated and often unique offering in a limited time period, they form an additional reason for cultural tourists to visit a place. They can cause a place to rise on the shortlist of places the tourist has in his or her mindset of attractive destinations. Festivals and events are both effective instruments in attracting first time visitors as well as repeat visitors due to the differential advantage they can offer."* (City Tourism and Culture, 2005 p. 44.)

According to recent surveys we can state that the majority of cultural tourists are motivated to take part in event and festival tourism as well, since 88% agreed on an internet questionnaire that cultural festivals and events are important reasons for cultural tourists to choose to specifically visit a place (City Tourism and Culture, 2005 p. 44.).

Of course entertainment (of events and festivals) as a motivation in tourism is really diverse to analyse but we can state that these events, festivals and parades mainly cover cultural thematics such as music festivals and events and all the other forms of fine arts festivals and events (of course we can highlight gastronomy, religion, folk, film, history etc. topics as well). The different festivals can contribute to the development of the given areas or regions and also promote the cognition of the local population or residents of an area.

A very important role of festivals and events that (however they usually produce a timely concentration in the high season in majority) they act against seasonality, since a vast amount of festivals and events are organised in the low season.

On the other hand if we take into consideration the size of the megaevents and the carrying capacity of places for instance – so the huge amount of the number of people visiting these places in a relatively short time – we have to stress that cultural events and festivals could have a seriously negative impacts on the environment and on the local population as well.

Creative culture, creative tourism

The term creative culture and creative tourism is more and more widely used in recent cultural tourism trends researches and analyses. UNESCO's (2006) working definition of creative tourism is the following: "travel directed towards an engaged and authentic experience, with participative learning in the arts, heritage, or special character of a place. It provides a connection with those who reside in this place and create this living culture." (http://ec.europa.eu/culture/keydocuments/doc/study_impact_cult_creativity_06_09.pdf)

The most recent trends of cultural tourism investigate more and more on the topic of creative tourism. Of course the link between creativity and culture is obvious and it is also natural that those people who are involved in creative industries (artists or professionals that are active in cultural/creative industries) will be in a way or other linked and connected to culture and cultural tourism as well. So as a driving force of the ever developing and diversifying cultural tourism *"culture is taken to encompass traditional cultural and artistic activities (performing arts, visual arts, cultural heritage and literature) as well as cultural industries (printed works, multimedia, the press, cinema, audiovisual and phonographic productions, craft, design and cultural tourism)."* (The Impact Of Culture On Creativity, 2009, p. 21.)

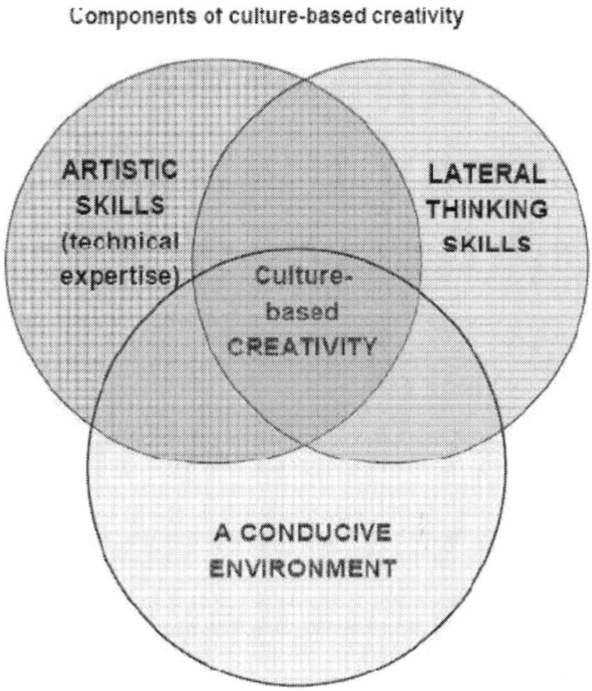

Components of culture-based creativity

Source: http://ec.europa.eu/culture/key-documents/doc/study_impact_cult_creativity_06_09.pdf

Figure. 8. Components of culture-based creativity

1.2 THE ROLE OF CULTURAL TOURISM IN THE GLOBAL TOURISM MARKET

In the 21st century the tourism global market creates an organic and interdependent system in which the supply and demand side experiences significant changes both in time and space and also from the perspectives of the quantitative and qualitative aspects or components. Newer and newer regions and tourism products will be involved in the international and domestic tourism trends as well and in the ever growing competition only such a tourism destination of tourism actor can survive which or who can provide an ever growing standard of quality.

RICHARDS (2009) states that *"Culture and tourism were two of the major growth industries of the 20th century, and towards the end of the century the combination of these two sectors into 'cultural tourism' had become one of the most desirable development options for countries and regions around the world."*

According to the recent changes of tourism trends it is obvious that visitors are more strongly involved in cultural activities than earlier although we have to highlight that the role of the 3S (or 4S as sun, sand, sea and sex) in mass tourism will still be (very) dominant. On the other hand as the new generations of visitors appear on the tourism market, now we can talk about a new 3S group or generation of tourists now mainly motivated by sport, spectacle and satisfaction. (Csapó & Matesz, 2007) Also we have to stress that one of the most important motivations for a tourism visit is getting (more and more and as diverse as possible experiences.

Some aspects of cultural tourism is summarized in the following table.

Table 8. Positive and negative effects/impacts of cultural tourism

Positive effects	Negative effects
• The development of the regional culture • Protection of the natural habitat • The accentuation of tourism regions • Strengthening of the local traditions and culture • Less seasonal, can extend the tourism season • Can be an important form of sustainable tourism	• Culture become commercialised • Destruction of the environment • Investments in tourism that act against the state of the environment • Architecture not characteristic to the local customs • Carrying capacity problems • Cultural tourism has only a dependent role (need for package) • Conflict source

Source: Based on HORVÁTH, 1999 own editing

If we take into consideration and observe the impact and importance of cultural tourism on the global tourism market we have to strongly emphasize that according to the recent research data published by the OECD in 2009 entitled The Impact of Culture on Tourism it seems that worldwide almost 360 million international tourism trips were generated by cultural tourism in 2007, accounting to around 40% of all global tourism (OECD, 2009). Furthermore if we take into consideration that these numbers were only directly affecting the tourism industry we have to stress that the indirect contribution of cultural tourism is naturally even higher due to its multiplicator effects. The mentioned study also stresses that the amount of money spent by a 'cultural tourist' is estimated to be as one third more on average than other tourists (Richards, 2009).

In this ever changing system of the tourism industry the role of cultural tourism is rapidly and constantly growing in the latter decades but we also have to highlight that the positions of the classic mass tourism often characterised by the 4S (sun, send, sea and sex) will be the most dominant form of tourism for a very long period of time (Figure 9).

The cultural tourism products will only be able to survive and attract more and more tourists – of course taking into consideration the basic principles of sustainable tourism – by applying an up-to-date and competitive cultural tourism product development approach which, according to a recent cultural and heritage tourism product research paper created in Ontario in 2009, is mainly based on quality, distinctiveness, economic benefit and creativity (Figure 10.).

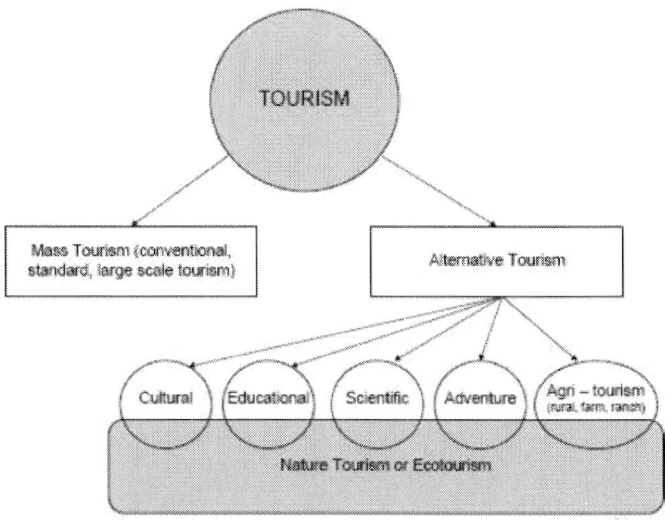

Source: MIECZKOWSKI, 1995 p. 459.

Figure 9. The role and place of cultural tourism within alternative tourism

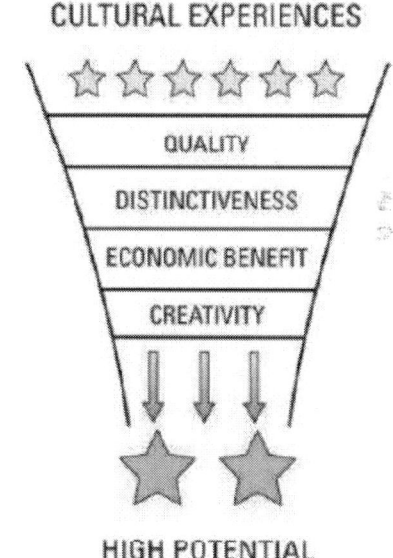

Source: Ontario Cultural and Heritage Tourism Product Research Paper, 2009.

Figure. 10. A 21st Century Framework for Evaluating Cultural Tourism Products

In accordance with the Ontario Cultural and Heritage Tourism Product Research Paper (2009) we believe that there are 5 key trends that will effectively and remarkably characterise the trends of cultural tourism in the near future.

These five elements are the new and emerging markets of the ever changing global tourism industry, the appearance and strengthening of the so called creative economy, agency and participation which is very much connected to the characteristics of the 'Y' generation and finally competition of excellence which will be a determining driving tool for cultural tourism product development. (Ontario Cultural and Heritage Tourism Product Research Paper, 2009)

When analysing cultural tourism product development we also have to be aware of the potential stakeholders which are forming the cultural tourism product. Understanding this process we should analyze Figure 11 which is demonstrating the major stakeholders in the formation of the cultural tourism product. This very complex system clearly shows that cultural tourism and its tourism product is dependent on many aspects from the individuals (either travellers or entrepreneurs), to local governments, NGO's and state organisations etc.

Finally when we try to analyse the cultural tourism product we have to analyse this issue from the statistical perspective as well. First of all we would like to stress that statistical data collection is very often country or location specific or even researcher specific. In this chapter we would like to introduce and present the statistical data collection method of the United Nations World Tourism Organisation which of course we consider to be the basic approach to this question.

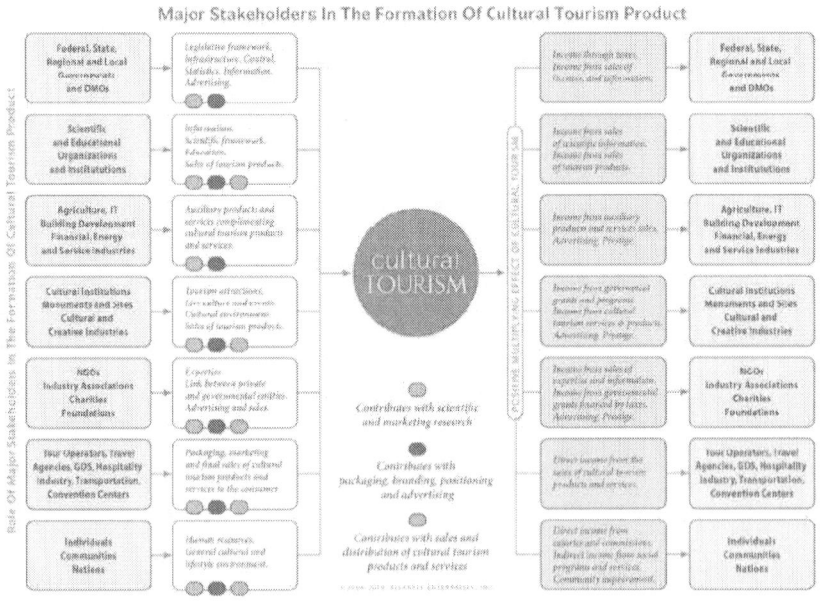

Source: http://www.culturalrealms.com/2009/07/stakeholders-role-in-tourism-and-the-effect-of-travel-industry-on-them.html

Figure 11. Major stakeholders in the formation of cultural tourism product

Table 9. List of categories of tourism characteristic consumption products and tourism characteristic activities (tourism industries)

Products	Activities
1. Accommodation services for visitors	1. Accommodation for visitors
2. Food and beverage serving services	2. Food and beverage serving activities
3. Railway passenger transport services	3. Railway passenger transport
4. Road passenger transport services	4. Road passenger transport
5. Water passenger transport services	5. Water passenger transport
6. Air passenger transport services	6. Air passenger transport
7. Transport equipment rental services	7. Transport equipment rental
8. Travel agencies and other reservation services	8. Travel agencies and other reservation services activities
9. Cultural services	**9. Cultural activities**
10. Sports and recreational services	10. Sports and recreational activities
11. Country-specific tourism characteristic goods	11. Retail trade of country-specific tourism characteristic goods
12. Country-specific tourism characteristic services	12. Other country-specific tourism characteristic activities

Source: International Recommendations for Tourism Statistics 2008 p. 42.

The publication of the 'International Recommendations for Tourism Statistics 2008' from the point of view of tourism statistical data collection provides a list of categories of tourism characteristic consumption products and tourism characteristic activities (tourism industries).

As we can see the 9th point of this elementary categorisation of tourism products and services deals with cultural tourism product and its services. In order to better understand this section of the tourism industry the mentioned research paper provides the more accurate and characteristic tourism statistical data background for cultural tourism in Annex 4 in the following:

Table 10. List of tourism characteristic products and grouping by main categories according to CPC (Central Product Classification) Version 2

9. Cultural services
96220 Performing arts event production and presentation services
96310 Services of performing artists
96411 Museum services except for historical sites and buildings
96412 Preservation services of historical sites and buildings
96421 Botanical and zoological garden services
96422 Nature reserve services including wildlife preservation services

Source: International Recommendations for Tourism Statistics 2008 p. 122.

4. CONCLUSION

Summarizing this chapter we have to state that the cultural tourism product and cultural tourism itself is a very complex segment of the tourism industry, both its demand and supply is diverse and versatile. Its future positions will most probably be strengthened directly and indirectly as well since with the change of the recreational needs of tourists and visitors the demand for cultural travels will rapidly grow as well (additionally when we consider the new appearing sending markets). Of course classic mass tourism will never considerably loose its market positions but the new tourists will have a more and more diversified need to get to know the different cultures and customs of the remote places

On the other hand we also have to take into consideration that the rapid growth and development of cultural tourism caused various aspects of new problems in the industry When analysing these recent trends we also have to stress that not only the needs of local communities has changed but also the motivations of the cultural tourists. According to this perception one of the most important international researches on this area the ATLAS research "has indicated that the experiences enjoyed most by cultural tourists tend to be those smallscale, less visited places that offer a taste of 'local' or 'authentic' culture. Tourists increasingly say that they want to experience local culture, to live like locals and to find out about the real identity of the places they visit." (Richards, 2009)

In the analysis of a tourism product we have to be aware not only the positive effects but the negative aspects of tourism development as well. The ever growing and rapidly increasing cultural tourism in the recent years has raised the question whether it really serves the needs of sustainable tourism especially in small communities. Cultural tourism started as a form of alternative tourism and nowadays it can be considered – in certain tourism destinations – as a dominant part of mass tourism.

Local communities have to face with the degradation of their 'original' culture so "there are a growing number of places in search of new forms of articulation between culture and tourism which can help to strengthen rather than water down local culture, which can raise the value accruing to local communities and improve the links between local creativity and tourism." (Richards, 2009) According to this new trend, it seems that one of the most important trend and development of cultural tourism in the recent years lead us to the establishment of creative tourism which serves the needs of a more sustainable cultural tourism in today's tourism industry. So based on the vulnerability of the destinations we strongly have to stress that only conscious tourism planning methods and practice will be able to take into consideration the principles of sustainability and carrying capacity in the given (cultural) tourism region.

REFERENCES

1.   ATLAS (2009). *ATLAS Cultural Tourism Research Project* http://www.tram-research.com/atlas/presentation.htm
2.   Arts Industry Tourism Council (1997). *Cultural Tourism Development in Victoria* http://www.culturaldata.gov.au/publications/ statistics_working_group/cultur al_tourism/cultural_tourism_statistics

3. Aubert, A. & Csapó, J. (2002). *Unique Features of the Tourist Attractions in Hungary's Historical Small Cities*. In: Settlement Dynamics and Its Spatial Impacts Siedlungsdynamik und Ihre Räumliche Wirkungen. Ed.: Aubert, A. & Csapó, J. University of Pécs Department of Tourism, Pécs. pp. 137-147.

4. Berki, M. – Csapó, J. (2008). *The Geographical Basis for the Development of Thematic Routes*. In. Progress in Geography in the European Capital of Culture 2010. University of Geography – University of Pécs. Geographica Pannonica Nova 3. Pécs. pp. 161-173.

5. Berki, M. (2004). *Az örökségturizmus szerepe és fejlesztési lehetőségei Pécs idegenforgalmában*.In: Turizmus Bulletin. 2004/1. pp. 41-48.

6. Bureau of Tourism Research (1998): *Cultural Tourism in Australia* Commonwealth of Australia. 80. p.http://www.ret.gov.au/tourism /Documents/tra/Snapshots%20and%20Factshe ets/OP27.pdf

7. Csapó, J. & Matesz, K. (2007). *A kulturális turizmus jelentősége és szerepe napjaink idegenforgalmában*. In: Földrajzi Értesítő. 56. évf. 3-4. füz./2007. pp. 291-301.

8. Cultural Tourism Industry Group, http://www.cultural tourismvictoria.com.au/

9. Medlik, S. (1996). *Dictionary of Travel, Tourism and Hospitality Terms*. Butterworth-Heinemann; 2 edition 1996 ISBN-10: 0750628642 | ISBN-13: 978-0750628648, 360 p.

10. Hofstede, G. (1997). *Cultures and Organizations: Software of the mind*. New York: McGraw Hill; 1 edition. ISBN-10: 0070293074 | ISBN-13: 978-0070293076. 279 p.

11. Horváth, A. (1999). *Turizmus a kultúrában: (kultúra a turizmusban)* Magyar Művelődési Intézet, Budapest. 47 p.

12. http://ec.europa.eu/culture/keydocuments/doc/study_impact_cult_creati vity_06_09.pdf

13. http://ec.europa.eu/culture/our-programmes-and-actions/doc2485_en.htm

14. http://ec.europa.eu/culture/our-programmes-and-actions/doc413 _en.htm

15. http://en.erih.net/

16. http://project.heritour.com/

17. http://whc.unesco.org/

18. http://whc.unesco.org/en/list/

19. http://whc.unesco.org/pg.cfm?cid=31&l=en&action=stat&&&mode=ta ble

20. http://whc.unesco.org/pg.cfm?cid=31&l=en&action=stat&&&mode=ta ble

21. http://www.culturalrealms.com/2009/07/stakeholders-role-in-tourism-and-the-effectof-travel-industry-on-them.html

22.   http://www.culture-routes.lu
23.   http://www.roshan-institute.org/474552
24.   http://www.sccs.swarthmore.edu/users/00/ckenned1/culture.html
25.   http://www.travelindustrydeals.com/news/5041
26.   http://www.unesco.org/en/cultural-diversity/heritage/*ICOMOS Charter on Cultural Tourism* (1976).
27.   http://www.icomos.org/tourism/tourism_charter.html
28.   *ICOMOS Charter for Cultural Tourism* (1997). http://www.icomos.org/tourism/
29.   McKercher, B. & Hilary C. (2002). *Cultural Tourism: The Partnership Between Tourism and Cultural Heritage Management* New York: Hayworth Hospitality Press, 2002 (262 p.)
30.   Michalkó, G. (1999). *A városi turizmus elmélete és gyakorlata* Magyar Tudományos Akadémia Földrajztudományi Kutató Intézet, Budapest. 168 p.
31.   Michalkó, G. (2004). *A turizmuselmélet alapjai* Székesfehérvár: Kodolányi János Főiskola, 2004. 218 p. (Turizmus akadémia; 1.)
32.   Michalkó, G. & Rátz, T. (2011). *Kulturális turizmus* Turisztikai terméktervezés és fejlesztés digitális tankönyv. TÁMOP TÁMOP-4.1.2-08/1/A-2009-0051. 20. p .
33.   Mieczkowski, Z. (1995). *Environmental Issues of Tourism and Recreation* Lanham, MD: University Press of America, Inc. ISBN-10: 0819199958 ISBN-13: 9780819199959
34.   *National Trust for Historic Preservation's Heritage Tourism Program* http://culturalheritagetourism.org/documents/2011CHTFactSheet6-11.pdf
35.   Nyíri, Zs. (2004). *Turistafogadás az egyházi helyszíneken* In: Turizmus Bulletin 2004/1. pp. 27-32.
36.   OECD (2009). *The Impact of Culture on Tourism* Paris: OECD 159 p. http://www.em.gov.lv/images/modules/items/OECD_Tourism_Culture.pdf
37.   Office of National Tourism (1997). *Fact Sheet No 10 Cultural Tourism* www.Dcita.gov.au/swg/publicationsculttour.html
38.   *Ontario Cultural and Heritage Tourism Product Research Paper* (2009). Queen's Printer for Ontario, 70. p. 231
39.   http://www.mtc.gov.on.ca/en/publications/Ontario_Cultural_and_Heritage_Tourism.pdf
40.   Puczkó, L. & Rátz, T. (2000). *Az attrakciótól az élményig. A látogatómenedzsment módszerei* Geomédia, Budapest. 399. p.
41.   Puczkó, L. & Rátz, T. (2007). *"Trailing Goethe, Humbert and Ulysses Tourism: Cultural Routes in Tourism"* Cultural Tourism: Global and Local Perspectives, Haworth Press, New York.

42.   Richards, G. (1996). *Cultural Tourism in Europe* CABI, Wallingford. Available to download from www.tram-research.com/atlas

43.   Richards, G. (Ed.) (2001). *Cultural Attractions and European Tourism* CAB International, Wallingford. http://igbraiutama.files. wordpress.com/2011/05/cultural-attraction-ineiurope.pdf

44.   Richards, G. (2005). *Cultural Tourism in Europe* The Association for Tourism and Leisure Education (ATLAS) www.atlas-euro.org 254 p.

45.   Richards, G. (2009). *Tourism development trajectories – From culture to creativity?* Tourism Research and Marketing, Barcelona. Paper presented to the Asia-Pacific Creativity Forum on Culture and Tourism, Jeju Island, Republic of Korea, 3-5 June 2009. http://www.tram-research.com/atlas/APC%20Paper%20Greg%20Richards.PDF

46.   Sanyal, A. (2009). http://anandasanyal.blogspot.com/2009/06/ethnic-tourism-is-travelmotivated-

47.   by.html

48.   Shackleford, P. (2001). *The social context of cultural tourism* KéK folyóirat, Budapest, pp. 29-41.

49.   Stebbins, R.A (1996). *Cultural Tourism as Serious Leisure* Annals of Tourism Research v23 n4: 948–950

50.   Swarbrooke, J. (1994). *The Future of the Past: Heritage Tourism In to the 21st Century* In: A.V. Seaton (ed.) Tourism, the state of the art, Wiley, Chichester, 1994. 222-9.

51.   *The Impact Of Culture On Creativity* (2009). A Study prepared for the European Commission (Directorate-General for Education and Culture) 240   p. http://ec.europa.eu/culture/keydocuments/doc/study_impact_cult_creati vity_06_09.pdf

52.   Timothy, D.J. & Boyd, S.W. (2003). *Heritage tourism* Pearson Education, 2003 - 327 p

53.   Trócsányi, A. & Csapó, J. (2002). *A zöld sziget – Turizmus egy kicsit másképpen* Mandulavirágzási tudományos napok, Pécs, 2002. március 4. In: Az irlandisztika nemzetközisége –Az ír kultúra, történelem, politikai és gazdasági élet kérdései összehasonlító megközelítésben. Hartvig G., Kurdi M. & Vöő G. (Eds.) pp. 10-18.

54.   Pécsi Tudományegyetem.

55.   Tylor, E.B. (1871). *Primitive Culture: Researches into the Development of Mythology, Philosophy, Religion, Language, Art and Custom* Harvard University. Boston, Estes & Lauriat. 491 p.

56.   *UNESCO's Convention Concerning The Protection Of The World Cultural And Natural Heritage* (1972). Paris, http://whc.unesco.org/en/conventiontext/

57.   UNWTO (2008). *International Recommendations for Tourism Statistics* Draft Compilation Guide Madrid, March 2011 Statistics and

Tourism Satellite Account Programme. 121 p. http://unstats.un.org/unsd/tradeserv/egts/CG/IRTS%20compilation%20guide%207%20march%202011%20-%20final.pdf

58.   World Tourism Organization and European Travel Commission (2005). *City Tourism &Culture – The European Experience* 137 p. www.etc-corporate.org

CHAPTER 4

Cultural Districts, Tourism and Sustainability

Giulio Maggiore[1] and Immacolata Vellecco[2]

[1]Unitelma Sapienza University

[2]Institute for Service Industry Research, National Research Council, Italy

1. INTRODUCTION

Tourism may be an important development opportunity for many regions, especially for those who do not have a solid industrial tradition but a good amount of cultural resources. These resources, in fact, can become the key attraction on which a tourist destination may be built, setting in motion a process that can offer an important contribute to the local community's well-being. But resources are not in themselves a guarantee of success. All the most important stakeholders have to be committed to the purpose and coordinated in order to make possible the achievement of this goal. A clear and shared vision must be supported by a strong collaborative network rich in social capital.

According to this perspective, literature on destination management has flourished in the last years, drawing attention to the importance of a systemic approach to organise and promote a territory as an attractive tourist product. This approach can effectively bring to good results in terms of increase in tourist flows, with all the consequent benefits to the local economies. Nevertheless, it is not always able to ensure a sustainable development path. The emphasis on the tourist success of the destination and on the immediate economic returns to the specialized firms in the region may induce to neglect the need for an effective valorisation of cultural resources. The tendency to exploit the resources may prevail against the attention to develop their deeper potentialities.

To avoid this risk, it may be useful to employ the concept of "cultural district", where *culture* is considered a potential source of attractions for tourists as well as an opportunity to enhance local human resources and increase creativity and innovation in local production systems. Since the nineties, many studies have used this term referring to local development models based on tourism and culture, but the discussion about the nature and the characteristics of cultural districts is still ongoing. A general ambiguity characterizes the debate on the subject, as different concepts are collected under the same label, thus concurring to create a lot of misunderstanding also as far as active policies for local development are concerned.

This work aims at giving a contribution to clarify the concept of cultural district as an innovative opportunity for sustainable development, on the one hand highlighting the differences with other similar models and, on the other hand, identifying its very distinctive features. A clearer understanding of the model and of its strengths can, in fact, help to define the mechanisms of value creation that it is able to activate.

The virtuous circle of a sustainable development may, actually, be realised only on condition that culture is considered not merely as a "product" to sell, but as a synergistic agent that provides all the local industries with production systems, operational contents, management tools, creative practices, symbolism and identity. In this way, culture can offer a fundamental contribution to enhance creative potential, identity and social capital, allowing the integration between various local businesses, in order to start and carry on the development of a diversified economy in the long term.

After having explored the various implications of the peculiar way cultural district can create value and spread it over the territory, we focus on the issue of the district "creation", which is essentially the main concern for policymakers. The problem is that districts cannot be created, as they are the result of a spontaneous process where local actors progressively develop a common vision, become aware of their territorial identity and discover their mutual interdependencies. Only a full immersion of local stakeholders in the "industrial atmosphere" and the accumulation of social capital can give policymakers a concrete chance of success. On the other hand, the cultural district requires also a strong governance by an authoritative leader, able to drive the change process and create the institutional conditions to facilitate the achievement of the mission.

The final part of the work aims at exploring the delicate role of public institutions, pointing out some possible policies and actions that can be carried out in order to foster the birth of a cultural district in the territory.

2. CULTURAL DISTRICTS AS A KEY FOR LOCAL DEVELOPMENT

2.1 The concept of sustainable development

Researches related to the issue of sustainability show that sustainable development is a complex and multidimensional concept, which combines efficiency, equity and intergenerational equity. Ciegis et al. (2009) stated that economic literature offers over 100 definitions of sustainable development and cited the work of Jacobs (1995) mentioning as many as 386.

The Brundtland Commission's brief definition of sustainable development as the "ability to make development sustainable – to ensure that it meets the needs of the present without compromising the ability of future generations to meet their own needs" is surely the most famous definition.

The concept includes two goals that, despite seeming to be contradictory, have to be achieved simultaneously (Ciegis et al., 2009):

- to ensure appropriate, secure, wealth life for all people (the goal of development);

- to live and work in accordance with bio-physical limits of the environment (the goal of sustainability).

As a general concept, sustainable development encompasses three fundamental approaches: economic, environmental, and social development, which are interrelated and complementary. These three "pillars" of sustainable development were indicated in the 2002 World Summit on Sustainable Development, marking a further extension of the standard definition.

The economic sustainability seeks to maximize the flow of income and consumption that could be generated while, at least, maintaining the stock of assets or capital and safeguarding its optimal amount for the future generations. The environmental approach pays attention to stability of biological and physical systems. According to this approach, the task of economic development is to determine the natural systems limits for various economic activities. Socio-cultural sustainability concept reflects the interface between development and dominant social norms and strives to maintain the stability of social and cultural systems and their ability to withstand as stocks. It also implies preservation of social capital and shared global responsibility for the planet, including corporate social responsibility.

It was also noticed (Kates et al., 2005) that, in the wide debate on sustainable development, the concept maintains a creative tension between a few core principles and an openness to reinterpretation and adaptation to different social and ecological contexts. The original emphasis on economic development and environmental protection has been broadened and deepened to include alternative notions of development (human and social) and alternative views of nature (anthropocentric versus ecocentric). Indeed, nature and environment can be valued for their intrinsic value or for its utility for human beings and as a source of services for the utilitarian life support of humankind. The concept of development, originally focusing on economic activities and productive sectors providing employment, desired consumption and wealth, extended its scope to human development and included an emphasis on values and goals, such as life expectancy, education, equity and opportunity; on the value of security and well-being of national states, regions, and institutions as well as the social capital of relationships and community ties.

A further aspect has been recently introduced (Helm, 1998) that is to say, the institutional/organisational aspect, due to the importance and significance of institutions in the policy. Effective and properly functioning institutions and institutional capital are essential for sustainable development in the realisation of the social, economic and environmental aims set by society.

Sustainable development is an overarching objective of the European Union launched in the EU Sustainable Development Strategy in Gothenburg in June 2001 and reaffirmed in the Renewed Sustainable Development Strategy (SDS), in June 2006.

If tourism is adequately addressed toward the path of sustainable development, it can represent an effective chance to promote economic growth, employment, social progress, as well as the protection and enhancement of cultural and environmental heritage. In addition, ensuring the economic, social and environmental sustainability of tourism is also crucial for the continued growth, competitiveness and commercial success of the industry itself. In this regard, in the Communication Agenda for a Sustainable and Competitive European Tourism, the European Commission provided all actors with some basic guidelines to create "the right balance between the welfare of tourists, the needs of the natural and cultural environment and the development and competitiveness of destinations and businesses" (Comm, 2007, 621). A long term planning and a continuous reporting are recommended, as well as an integrated and holistic policy approach where all stakeholders share the same objectives.

As real-world experience has shown, however, achieving agreement on sustainability values, goals, and actions is often difficult and painful work, as different stakeholders values are forced to the surface, compared and contrasted, criticized and debated (Kates et al., 2005).

The role of culture in local development

Culture may play a main role in supporting sustainable development processes, not only because it can provide some important tourist attractions which can help to enhance the competitiveness of a territory, but also because of its social implications. Indeed, cultural development is generally considered to be an essential part of social development, and cultural diversity provides sources for creative expression that are increasingly being harnessed by players in the creative industries.

One definition of ɔbcultureɒ given by Throsby (2001) refers to the set of attitudes, practices and beliefs that are fundamental to the functioning of different societies and groups defined in geographical, political, religious, or ethnical terms. Culture thus finds its expression in a particular society's values and customs, which evolve over time as they are transmitted from one generation to the next. Accordingly, culture is both tangible and intangible. The stock of tangible cultural capital assets consists of buildings, structures, sites and locations endowed with cultural significance and artworks and artifacts existing as private goods, such as paintings, sculptures, and other objects. Intangible cultural assets includes the set of ideas, practices, beliefs, traditions and values which serve to identify and bind a given group of people together, however the group may be determined, together with the stock of artwork existing as public goods in the public domain, such as certain instances of literature and music.

Several recent studies emphasize that the field of culture and creativity is launching a much needed boost to economic activity and employment dynamic in all advanced economies (KEA, 2006). Moreover, it seems to be at the basis of the exponential growth processes that Asian economies have being registered over the last decades (Yusuf & Nabeshima, 2005).

Research has also emphasised the potential of these industries in developing countries (UNCTAD, 2004). Creativity, more than labour and capital, or even traditional technologies, is deeply embedded in every country's cultural context. Excellence in artistic expression, abundance of talent, and openness to new influences and experimentation are not the privilege of rich countries. With effective nurturing, these sources of creativity can open up new opportunities for developing countries to increase their shares of world trade and to "leap-frog" into new areas of wealth creation. Because the marriage of technological application and intellectual capital provides the main source of wealth in this sector, continuous learning and a high degree of experimentation are key to achieving sustained and cumulative growth. This mixture can produce very fast growth.

The cultural sector is a powerful driver of development, through the attraction of businesses and talented people (Florida, 2002) . This concept focuses on a further dimension of culture, as location factor in attracting skilled and creative people and promoting social cohesion, which can stimulate or increase the dynamics of personal and business networks.

So culture makes the difference and can produce important opportunities for local development, as the cultural paradigm is the tool which re-defines the significance of place in terms of identity, territoriality and functionality (Battaglia & Tremblay, 2011). This result can be achieved only if the territory is able to find effective forms of self-organisation which enact the virtuous circle of a local development based on cultural resources. Recent literature agrees that the model of cultural district can be a good answer to this problem.

Cultural district

Cultural district emerges as an innovative model of sustainable development where renewable resources as culture have been assigned a role in the production of income and employment. In a general meaning, cultural district is a conceptual model to build up a local development strategy based on the cultural dimension and inspired by the logic of sustainability.

Cultural district is an innovative concept for land use planning based on the set of values that characterise the local identity and transform cultural heritage into a tool for the community development, targeting decisions, planning, investments. The recognition of belonging to a specific local culture is a prerequisite to enable networks and projects, within the cultural sector, aiming to harness the potential hidden in the territory (Carta, 2002, 2004).

Culture enhancement can be driven by cultural managers in traditional ways, nevertheless contributing to creation and development of other productive activities of the cultural sector: such as research, cataloguing, custody, implementation of educational practices, exhibition activities (Valentino, 1999). In a broader sense, the enhancement of cultural resources is conceived as a tool to attract tourists in a territory and, as a consequence, to increase the demand for goods and services that caters to the local market. In this case, the enhancement process works as a policy measure; it involves a greater number of stakeholders and economic activities, but requires a context of higher quality (architectural and relating to landscape, but also social) as well as an adequate supply of hospitality services and infrastructure.

The artistic and cultural heritage is the factor of production enabling the formation of a cultural district as *High Cultural Local System* (Lazzeretti, 2001) and its enhancement is an investment to create network and learning economies. Art City is considered (Lazzeretti, 2001) as "an incubator of new entrepreneurship, as the link between economic and social community, as the connector between the different cultures " and then "as a possible form of organisation of socio-economic-territorial space within which different sectors are characterised by spatial proximity and cultural organisation of work" where research has to focus issues of governance and identify key players, measuring the density and discovering the shapes of the networks.

The identification of urban cultural district as a cluster of buildings and spaces dedicated to arts, cultural services and production of goods based on culture (Santagata, 2005) is also reflected in a large field of studies pointing on the use of artistic and cultural services to tackle the industrial and economic decline and draw a new image of a city capable of attracting visitors and leveraging on tourism to boost the local economy.

In this case, the district is also the result of an urban planning favourable to the arts and cultural activities (museums, libraries, theatres, art galleries, concert halls), enabling industrial activities based on culture (film studios, rooms music recording, television stations), as well as activities traditionally addressed to the attraction and reception of visitors and tourists (restaurants, bars, gift shops and gift items, high quality clothing). Cultural quarters and creative quarters[1] are the product of interactions between urbanisation, culture and creativity, especially if we pay attention to the role of networking activities and clustering processes in specific urban areas. A regeneration process based on cultural quarters can be significant and has to be supported by an official objective of development, regarding social and economic concentration of actors which are interested in boosting culture and creativity, and their impact within local contexts (Landry, 2000; Santagata, 2002; Roodhouse, 2006). They increase the strategy of regeneration and renewal of complex of buildings and of depressed urban areas, supporting social inclusion and territorial cohesion as the main driving forces of socio-territorial innovation processes (Tremblay et al., 2009).

Local development arises in a new urban landscape made by powerful regional economies based on the city, where creativity and cultural production play an essential role in sustaining economic growth. Within this context, the importance of the proximity of individuals emerges, enabling the human act to produce creative thoughts and innovation (Bucci & Segre, 2009). By allowing the knowledge of one individual to spill over onto others, the productivity of the others is improved in a virtuous circle. Furthermore, the widespread diffusion of knowledge derived from knowledge spillovers enhances productivity not only among individuals working within the same sector, but also across different and sometimes apparently very distant sectors, creating a process of cross-fertilization.

The usefulness of cultural district regardless of its ability to generate profit for itself is affirmed by Sacco and Pedrini (2003), who stated that this model has value and meaning because of its ability to complement other sectors of the local system, resulting in innovative synergies otherwise unattainable. The competitive ability is linked even more to orientation towards innovation; so, culture is assuming an increasingly strategic role as a synergistic agent that provides other sectors of the production system with contents, tools, creative practices, value added in terms of symbolic value and identity. That induces many local systems to invest more heavily in offering not only culture, but also in allowing a deeper integration between culture and the various aspects of social everyday life.

Many cities have sought to create cultural districts, directed primarily at attracting suburbanites, tourists, and conventioneers. But most cities already have cultural districts, neighbourhood-based cultural clusters that have emerged without planning or massive public investment. What is more—because they are complex ecosystems that combine artistic production and consumption and a mix of institutional forms, disciplines, and sizes—they have a degree of sustainability that a planned cultural district is unlikely to match (Stern & Seifert, 2007). Recognizing the importance of natural cultural districts to the metropolitan arts world turns our understanding of cultural planning and policy on its head. The goal of policy and planning should be to nurture grass-roots districts, remove impediments that prevent them from achieving their potential, and provide the resources they need to flourish.

The identification and involvement of key stakeholders was also identified as pivotal; consensus-based decision making importance has been widely recognised also in the tourism literature (Bramwell & Sharman, 1999; Jamal & Getz, 1995) as well as the role of information sharing for the attainment of both short and long term objectives.

Anyway, the concept of cultural district is very wide and ambiguous, as it covers a wide range of different meanings which reflects its multidisciplinary origins and its heterogeneous practical applications. A possible classification distinguishes five typologies based on the different cultural resources which play the focus role within the district (Santagata, 2005):

- industrial cultural districts of material culture, based on the production of goods and   services of material culture, enhanced through a wise use of institutional rules;

- museum cultural systems, usually located in the historical urban downtown and based on a process of urban planning act to enhance the artistic and historical heritage by using an innovative network capable of producing a very strong collective image (brand);

- tourist cultural districts, characterised by the supply of traditional cultural services (heritage, folklore, museums, spas), a high concentration of hotels and hospitality- related activities and a local production of craft art and material culture;

- cultural heritage systems, taking the form of a circuit or network that connects individual sites or monuments, characterised by a common identity, reinforced by the production of collective services;

- urban cultural districts, also known as "American Cultural Districts", that is to say a cluster of buildings and spaces dedicated to arts, cultural services and the production of goods based on culture, aiming at revitalizing declining urban areas by developing artistic and cultural services.

This distinction shows that the term "cultural district" in a broad sense may become an "umbrella expression" where many different kinds of local clusters of organisations can be included, from the classical industrial district up to the more recent forms of metropolitan quarters. This "ecumenical" approach can belittle the interpretative value of the concept of cultural district, as "in the night all cows are black".

A narrower definition of the concept, which emphasises its differences with the other kinds of districts (specifically industrial and tourist districts) may help to highlight its peculiar characteristics that make it an innovative model capable of showing new and original development opportunities.

3. SIMILARITIES AND DIFFERENCES AMONG CULTURAL, INDUSTRIAL AND TOURIST DISTRICTS

The term "district" is a very fashionable label used to categorise several successful economic experiences based on the aggregation of many small enterprises sharing the same geographical space of action. The expression "industrial district" - first used by Alfred Marshall in his *Principles of Economics* (original edition in 1890) – was rediscovered by Italian industrial economists in the 1980s to explain the success of Italian SMEs located in some dynamic regions in the North of the country (Becattini, 1987, 1989; Bellandi, 1982, Dei Ottati, 1986; Brusco, 1989). The district model perspective outlined the importance of intangible values, such as "industrial atmosphere" or tacit knowledge sharing, as key factors to foster the competitiveness of all the companies included in the local cluster. This became a model even for other countries, to find an alternative to the tradition of Fordism and mass- production (Piore & Sabel, 1984).

Despite the fact that globalisation has recently cast many doubts on the fitness of the model for the new challenges of international competition (Varaldo, 2004), the success of the term has been proven by its diffusion in other economic industries and contexts, so that now we have "technological districts", "tourist districts" and "cultural districts", just to mention the more common labels derived from the original conception of "industrial districts". In fact, the semantic ambiguity caused by the abuse of this terminology is very high, as the same words are often applied to describe very different conditions.

We have just described the large variety of different theoretical understandings of the concept of "cultural districts" and we can say that the situation is not less intricate with regard to the other typologies. Nevertheless, it is possible to point out some features of these models which can contribute to a clearer definition of the peculiarities of each of them. In particular, the concept of "cultural district" may be enlightened by a comparison with the other closest ones, that are those of "industrial" and "tourist" districts[2]. This comparison does not aim to entrap the theoretical fluidity of this issue into a rigid framework; on the contrary, the purpose is to use some ideas commonly shared by scholars in order to reduce the space of ambiguity and gain a better understanding of the phenomenon.

The features common to the three models, which justify the use of the same label of "district", are those referable to the Marshallian theory (Belussi, Caldari, 2009) which represents the main reference for the majority of scholars in the field. These factors allow the network externalities which are the main distinctive characteristics of districts:

- the role of "industrial atmosphere" as a cultural glue, able to put together the economic and social actors of the local community;

- the presence of a qualified and specialized workforce;

- the free circulation of tacit knowledge;

- the sharing of common values;

- the proximity of complementary companies;

- the mutual trust among local people.

If these factors are common to every kind of district, it may be useful to distinguish other features among the three typologies we are comparing. These are:

- the *"catalyst" of the district*, that is the component capable of activating local resources, combining them with each other in such a way as to make possible the development of the network externalities which, as we have seen, are the lifeblood of any district;

- the *role played by territory*, that can be expressed with a "metaphor" in order to synthesize its function towards the district;

- the *"mission" of the district*, that is the reason why it exists, according to the point of view of its economic actors;

- the *model of governance*, that is the mix of solutions adopted to coordinate the strategies and the actions of the local actors.

It is possible to point out the differences among industrial districts, tourist districts and cultural districts by focusing on these four features.

Industrial district

In the case of industrial districts, the role of **catalyst** is played by a specific manufacturing activity in which local businesses develop a meaningful and productive specialisation. It becomes increasingly part of a tradition that involves all local stakeholders: firms, public institutions, non-profit organisations, training agencies, professionals, individual residents. Everyone contributes to consolidate and develop the system of skills, facilities, infrastructures which innervate the district, determining its identity. In this case **territory** is experienced as a "factory", an open and fragmented space, where production lines are replaced by a network of small and independent businesses, while many small firms, serviced by a few skilled workers, take the place of crowded manufacturing plants. This perspective reflects the thought of Marshall, who theorized the district as a mode of organising production alternative to the large enterprise, where network externalities are used to compensate for the loss of economies of scale.

The *mission* of the district is the competitiveness of local enterprises, which is pursued through different levels of awareness by the various stakeholders acting in the system. Entrepreneurs aim to reinforce the competitiveness of their own enterprises, but the mutual interdependencies existing within the network make it clear that each firm can be more competitive only if the same happens to the other complementary firms belonging to the same district. The same can be said even for other stakeholders, such as public institutions, educational organisations or nonprofit associations, which tend to give a particular attention to the needs and requirements expressed by local enterprises, as they know that the well-being of the local community depends on the success of the industrial district. The system relies on the action of an "invisible hand" that binds individual to collective interests into a unique network of interdependencies.

The *model of governance* of the industrial districts is generally based on a tendency to spontaneous coordination, typical of polycentric networks lacking in a leader subject.

Informal relationships and rules, often implicit, ensure the proper functioning of the system[3]. Even if there is the emergence of a leader - usually a more dynamic and competitive enterprise - the district still tends to rely on traditional spontaneous coordination mechanisms (Lazerson & Lorenzoni, 1999), while local institutions tend to have a secondary role, aiming at facilitating the dynamics of the district rather than driving them, according to a model of "heterarchical" governance (Sacchetti & Tomlinson, 2009).

Tourist district

In the tourist districts the role of *catalyst* is carried out by "destination", where this term is intended to mean more than territory itself[4]. The destination is, in fact, a physical space, but also a "mental space", which corresponds to the image of the area as perceived by its stakeholders (first of all, the visitors, but also all the operators and the residents themselves). The perceived image of the destination becomes the point of reference for the efforts of the actors in the district, all committed to consolidating and fostering this perception. So, for instance, if the cultural district has taken root, all restaurateurs, hoteliers and operators will adopt behaviors and attitudes consistent with the destination image, aware that the success of their companies depends on that of the whole territory.

Here *territory* acquires the connotations of "product", as it is not only the location where the production is organised, but also the heart of the supply system. While in the industrial districts territory is just a place of production, ignored by the majority of those who use the manufactured goods produced in that place, in tourist district there is the physical and temporal coincidence of production and consumption. Thus, territory changes its function: consumers become part of the system and play a main role within the process of integration which produces the network externalities. Territory is not a simple back-office for production activities, but the focal point where the "moment of truth" takes shape (Normann, 1984), thus becoming the core issue for the destination marketing mix.

Therefore, the *mission* of tourist district is the increase in the flow of tourists, which is the vital condition of any possible development process. All the stakeholders in the territory are, in fact, focused on providing services which can enhance the capacity to satisfy the tourist demand. Everyone is important, as the overall experience of a visitor is determined by the combination of a large amount of little events occurring during the visit. Everybody and every situation he meets during his experience, may offer a positive or negative contribution to his perception, influencing his level of customer satisfaction. The stakeholders must cooperate to deliver an effective response to user requests, even sacrificing their immediate interests to contribute to the overall competitiveness of the destination. This is the only effective way to guarantee the attractiveness of the territory in order to increase the tourist traffic, with clear benefits for the local economic system.

Concerning the need for coordination, it is adopted a *model of governance* based on the leadership of a subject which assumes the responsibility of guiding the system, as the "product territory" must be organised and put at the center of an effective marketing mix, oriented to the expectations of a well-defined target market. This role of leadership can be taken by public or private bodies, including aggregations of operators, as in the case of consortia. The leader which acts as a "Destination Management Organisation" (DMO) can have a more or less strong relationship with other subjects of the territory, according to different contexts and institutional arrangements that define specific powers, responsibilities and limits of delegation. Usually the effectiveness of a DMO's action depends on its legitimacy: if it is accepted by the majority of the local community members, it can do a good work. Otherwise, its efforts risk to be vain, as demonstrated by the failure of the attempts to impose a subject intended to fill this role, without creating the conditions to sustain its legitimacy.

3.3 Cultural district

In cultural districts the *catalyst* of local development processes is "culture" itself. The enhancement of local cultural resources are considered as a basic element of any dynamic evolution of the territory, while the success of local industries and tourist activities may be seen as a possible consequence of culture. The difference may seem very blurred in practice, but it has a decisive impact on the criteria adopted to establish priorities and define economic policies. Putting culture at the heart of the development model means accepting that times are those of culture, rooted in past centuries, with processes of change which can take decades to achieve visible results. It also means not to focus on a single aspect that reveals a side of that specific culture (a product or a tourist attraction), but try to get all the possible dimensions, exploring new fields of application and new forms of cultural manifestations. It is much more than just exploiting local resources: it deals with using culture as a great opportunity to support durable and sustainable process of growth, which combines economic, social and environmental benefits.

The role of *territory* may be effectively expressed with the metaphor of "source". Indeed, it is not just a region fit for a particular industry or a destination for incoming tourists, but a source of values and opportunities that can foster different developmental processes. The fundamental difference is that a dynamic view prevails, where territory is not only a physical place, but a "space of possibilities", which can evolve in multiple directions depending on a dialectical relationship with the people who inhabit it. Success depends on the richness and abundance of "source territory", but also on the ability to address the potential energy residing in the cultural resources towards effective purposes, by activating virtuous circles in the local social system.

In this context, the *mission* of the district is the increase in value of cultural resources, where the expression "increase in value" should be considered as something more complex than just using tourist attractions to generate touristic flows (as in tourist districts) or leveraging local competences to support competitive enterprises (as in industrial districts). It is, instead, a process that can include all these factors, but it can and should go much further, up to promote indirect effects, involving the activation of creative resources of the territory and, generally speaking, determining rise in quality of life.

In cultural districts the *model of governance* is generally "hybrid", as different conditions and styles of leadership coexist. The resources to be involved in the development process are, in fact, usually managed by different parties, in both the public and the private sector. On the public side, there is often a mix of overlapping responsibilities involving different authorities with territorial or specialised competences, often in a conflictual relationship, as they pursue different objectives (local development, protection and conservation, promotion, etc.). On the private side, there are different organisations, belonging to the profit or nonprofit sector, which give contributions to local culture and have a specific interest in taking part to the decision-making process. This situation requires the recourse to models of "public governance", where the institutions are called to trigger a virtuous relationship with other regional actors, overcoming the reasons for conflict and enhancing the initiatives to meet the expectations of all the stakeholders involved.

3.4 The originality of cultural districts

The proposed considerations allow us to draw an overall picture, sketching a conceptual positioning map of the three models, where the cultural district seems to be located in an intermediate position between the industrial district and tourist district (Fig. 1).

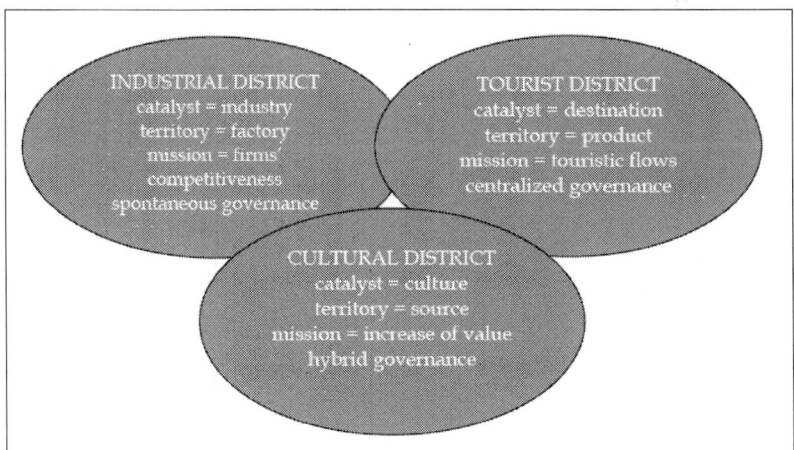

Figure1. Conceptual positioning of the models of industrial, tourist and cultural districts

While the latter, in fact, reveal a clearer identity , which emerges from the different focus in terms of catalyst, vocation of the territory, mission and governance models, cultural district seems to be characterised by a more nuanced profile, consisting of elements drawn from both, in varying combinations, which may reflect the peculiarities of the territories. In some cases, the district may take cultural characteristics closer to those typical of the industrial district, as happens for the "material cultures of the districts", while, in other cases, it will tend to converge toward the model of the tourist districts, when it is developed around highly attractive resources (Santagata, 2005).

This ambivalence does not mean that cultural district is just a "variant" of the other two models, without a specific identity: in this case, it could be considered as an unnecessary complication, which adds little to the understanding of the phenomenon. On the contrary, it has a specific profile that can be effectively outlined through the reference to the central role of culture as a catalyst of the local system. Culture is more important than immediate industrial or touristic success of the territory, as local stakeholders accept to invest on an intangible asset which can become a source of opportunities in the medium term. To escape the risk of trivializing the concept and emphasise its full sense of originality, some authors have added the word "evolved" after "cultural district", marking thus a clear distance from those who tend to provide a narrower perspective (Sacco & Pedrini, 2003).

According to this view, cultural district is an original socio-economic model for local development which shares some elements with the other two types of districts (industrial atmosphere, informal relations between SMEs, spontaneous circulation of knowledge, sharing of values rooted in the territory, etc..), but, at the same time, it is based on a different vision about the process of value creation related to the resources of territory. The basic assumption of the model is, in fact, that the value generated by the local "cultural resources" is not only connected to their immediate economic impact originated by the local typical product sold or by the money spent by tourists during their visits, as there are other sources of value, connected to possible derivates of culture, such as individual liberty, innovation, creativity or quality of life, which can support processes of growth perhaps less fast but usually more sustainable.

4. THE PROCESS OF VALUE CREATION BASED ON CULTURAL RESOURCES

The analysis of the dynamics of value creation based on cultural resources allows us to fully understand the specific features of cultural district model , as described in this work. In particular, it can help to highlight the differences with respect to the model of the tourist district, which is the main point of comparison, given that we intend to evaluate the contribution that cultural district can offer to a perspective of sustainable tourism.

To fully understand the dynamics of economic and social characteristics of cultural district, it is necessary to grasp the relationship existing between culture and value in all its nuances: this is a complex time and space relationship that takes years to express its most significant effects, often escaping attempts to quantify them.

The first factor of complexity is, of course, the fact that institutions and cultural activities have direct impacts at different levels: cultural, social, economic, fiscal, employment, environmental, real estate. As a result, to evaluate the effects of a resource or initiative in the field of culture, we should provide ourselves with different interpretations and tools, capable of measuring the effects in all the fields, even in contexts where it is difficult and questionable any attempt to quantify.

Moreover, it should not be overlooked that very different economic activities can be considered as "cultural", from the organisation of a music festival to the management of an archaeological site or a museum, from the provision of tourist routes to the preparation of a library, up to the staging of theater shows. It is obvious, as each type of cultural activity can generate different dynamics of value creation.

However, the typical perspective of cultural districts, which tends to a comprehensive interpretation of cultural resources and territory, highlights the limitations of a reductive approach, focused on the analysis of economic flows that relate to an individual asset or a single cultural initiative. Instead, it appears better to expand the scope of the analysis, by adopting a broader concept of "economic impact", which - going beyond the boundaries of the individual organisation or initiative - extends to all the economic effects arising from the presence of a group of sectors, companies or cultural institutions. In the latter sense, the focus of the analysis is mainly on the quantification of the "contribution" rather than the quantification of the "'impact" of culture in terms of production, employment, exports, etc. (Throsby, 2004).

A large literature has analysed the economic impact assessment produced by cultural heritage in a specific territory, pointing out four main effects:

- the generation of permanent (i.e. museums) or temporary (i.e. exhibitions and festivals) employment;

- the generation of revenue for companies belonging to the supply chain of services related to culture and heritage (protection, conservation, fruition) and for their suppliers of products and services (office furniture, security devices, hardware, software, storage products, construction materials and services, audio guides, merchandising, etc.).

- the attraction of tourism-related institutional initiatives and other cultural activities, which may function as attractions in themselves, calling tourists even during the low season and improving the image of the territory;

- the attraction of public investment, due to the presence of significant cultural resources that gain more attention from policymakers, encouraging the concentration in the territory of funding for the creation of infrastructure and the start of local development projects, with benefits disseminated to all stakeholders.

If we limit the analysis to economic effects, generally there are three levels of impact[5]: those of *direct spending, indirect spending* and *induced spending*. This is the basis for the assessment of the value created by culture, but it is not enough. Actually, if we completely accept the perspective of cultural district in its "evolved" meaning, we should also focus on the size of the social impacts of cultural activities, which do not produce immediate economic results, but can trigger processes for development in the medium to long term and make a most significant contribution to lasting value creation.

Many are, in fact, the non-economic benefits linked to the cultural heritage of a territory: the education of young people, the strengthening of the identity processes, the inclusion of disadvantaged social groups or minorities and immigrants, the development of a culture tolerance and human dignity based on the knowledge and protection of cultural diversity (EU, 2007). Culture is also a means of social re-integration or inclusion, because it gives people the opportunity to initiate and carry out their new projects and acquire new skills that restore confidence and self-esteem. Culture nourishes the human personality; it is the basis of educational processes and enriches the endowment - concepts, images, information, emotions - available to individual and community, thus facilitating reasoning, logic and semantic associations, analogies and contamination and providing people with more opportunities and a general ability to find solutions to problems as well as a flexible attitude in dealing with the "new".

These processes enacted by culture can create value for individuals, organisations and territories, due to the virtuous interactive connections which can be established between culture and creativity. The impact of culture on creativity and, indirectly, on the potential of economic and social innovation of a local community has been explored by recent studies which have highlighted all its many implications (KEA, 2009). According to these studies, culture can offer a crucial contribution to the development of new products and services, (including public services), driving technological innovation, stimulating research, optimising human resources, branding and communicating values, inspiring people to learn and building communities[6]. In other words, it is a key resource for the competitiveness of a territory. Furthermore, the presence of cultural amenities can contribute to attract creative talents, who, once gathered in a specific place, will create synergies and fruitful collaborations, thereby fostering further creativity (Florida, 2002).

In order to reflect this important opportunity of value creation specific to cultural district, the analysis of economic impact must go beyond the levels associated with the direct, indirect and induced effects, typically considered in the literature, to embrace an additional "layer" of benefits, which reflects the process of development of the area triggered by cultural resources through the power of creative potential, local identity and social capital. This fourth level of economic impact can be described as the "spread value ". It is very difficult to detect immediately, as it does not generate clearly identifiable expenses as economic benefits for local actors, but it builds up through long-term processes, gradually spreading and consolidating in the territory, turning out to be a decisive asset for local development opportunities (Fig. 2).

The focus on the "spread value" is decisive to understand the prescriptive relevance of cultural district model, whose utility relies on its capacity to "use" culture as an opportunity to produce development. Usually, investments on culture cannot be justified by their immediate economic returns, but the consideration on the "spread value" of culture may change the perspective, as it could prove that this kind of investments is affordable even according to an economic rationale, at least in a long-term view.

This is particularly true if we consider the purpose of sustainability with regard to tourist destinations. An excessive emphasis on the immediate economic returns connected to the first three layers of the value creation model may induce to stimulate flows of visitors even by exploiting local cultural resources. These are perceived as mere attractors of tourist interests and not as sources of creativity which must be integrated in complex processes of interaction with the local community to unveil all their potential of "spread" value creation. Consequently, they are treated as dead objects that belong to the past and are presented to visitors as a quality pieces of an "open air" museum, which have nothing to do with the present or with the future. This approach may produce good outcomes in the short, as the local community may take advantage of the economic effects of tourist flows, but it can be very risky in the long term, as the cultural assets of the region are not renewed and may be reduced by an excessive exploitation.

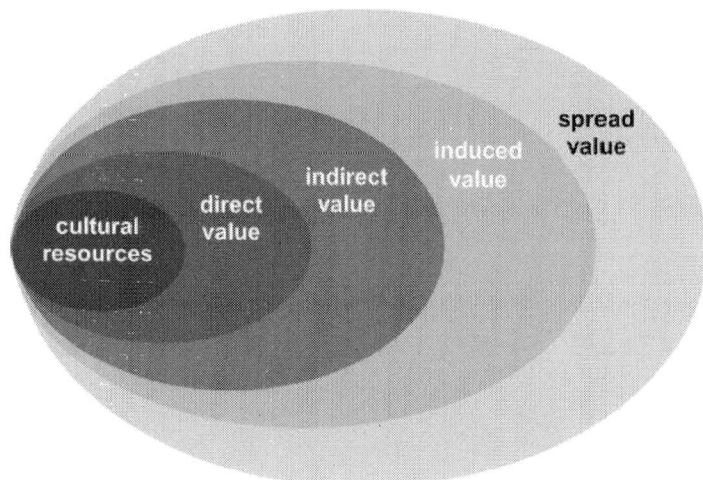

Figure 2. The process of value creation enacted by cultural resources

The model of cultural district suggests a very different way to draw development paths based on the resources of the territory. It considers culture as a vital value, which must be put in the center of social and economic processes, but not just as a tourist attraction. It has to become much more: a real engine for local society, capable to mobilize the best energies of the community in order to support a sustainable development process. Tourists are important, but citizens too. Culture is a "stock" of historical resources accumulated in the past centuries, but it is also a "flow" of new resources which can become tomorrow's stock.

5. CREATING A CULTURAL DISTRICT: THE REQUIREMENTS PUZZLE

Even if you agree with the idea that cultural districts may represent a useful chance to support sustainable and durable development, especially in areas that are poor in terms of economic resources but rich in terms of cultural heritage, nevertheless the passage from this belief to the effective "creation" of a cultural district is not an easy step. As a matter of fact, it requires a long term perspective and a strong commitment of all the main stakeholders in the territory, but the most complex issue is the search for a balanced mix between top-down planning and bottom-up spontaneous inventiveness. As stated before, the development of a cultural district requires a "hybrid governance", where a clear vision of a leader, responsible for planning, coordinating, stirring the local initiatives within a coherent framework, must coexist with a pluralistic and unstructured network of projects and casual actions, activated by local stakeholders.

Top-down programs aimed at creating districts where the "ground" is not ready have no hope of success, as well as the expectation of a completely spontaneous development, which may turn out to be a frustrating experience. A cultural district may emerge only from a process where an inspired strategy meets the interests and intentions of the most significant actors of the local community, planting its roots in a solid background of traditions and cultural assets. Therefore, those public administrators committed to the start-up of an experiment on a cultural district must focus their attention on facilitating the conditions for its development rather than on wasting time in accurate planning efforts which leave the community out.

Local policies should try to create these conditions which compose the ideal "humus" for the development of a cultural district. Sacco and Ferilli (2006) identify ten of these policies (which are also functional characteristics of the system): 1) Quality of the cultural offer;

2) Capacitation and training of the local community; 3) Entrepreneurial development;

4) Attraction of external companies; 5) Attraction of talent (artists and intellectuals);

6) Management of social and marginalization problems ; 7) Development of local talent;

8) Participation of citizens and local community; 9) Quality of local governance; 10) Quality of knowledge production. It is a good - even though not exhaustive - list of ingredients but not an ultimate recipe, as an ultimate recipe does not exist. Actually, every policy must be driven by a deep knowledge of the specific territory and of the dynamics which can help to convert cultural assets into creative processes in order to support sustainable development.

In a general vision of the functioning of cultural districts, the creative value chain starts from the cultural and artistic dimension and, then, drives economic systems into the field of applied research and creative production. Within this context, the pure cultural artistic dimension of the district and the creativity diffusion process which arises from it represent the key explaining factors of culture-led economic development. An effective strategy has to encourage investment in human and financial resources to prepare individuals to meet the challenges of the rapidly evolving post-industrial, knowledge-based economy and society. At the heart of this effort there is the identification of the vital linkage between art, culture and economic systems: the diffusion of knowledge is greatly influenced by cultural production, which originates in socially and economically embedded creative processes.

Some key elements are considered as fundamental to activate development dynamics:

- research and discovery of a shared social identity based on culture;

- production of innovation, knowledge and human capital through educational experiences, formal and informal networks, projects;

- dissemination of knowledge and cross-fertilization among productive sectors, in order to develop diversified economic activities and generation of new entrepreneurship;

- a view of the development of the territory as the ability to reach sustainable economic and social performances;

- strategic planning, with a strong involvement of local stakeholders.

The common cultural identity becomes the prerequisite for building specific development strategies for the territory based on the cultural dimension and inspired by the logic of sustainability. Social identity is one of the positive externalities associated with the processes of valorisation of cultural heritage, together with the production of research, innovation and knowledge which, if exploited in the area, through appropriate scale integration processes, increase the value produced by the region.

The enhancement of cultural assets targeting local stakeholders enables the recognition and strengthening of the local cultural identity. This is a set of values to rediscover and strengthen: they are related to the structure and features of the tangible and intangible cultural heritage and they also depend on social traits of identity, in terms of participation and empowerment of the community as well as educational experiences, information networks, sustainable development demands: the district is, therefore, the "future project" the local community aims to achieve by policy makers.

Local networks are unanimously recognized as a basic element and condition of possibility for the realization of the effective district, even when the main factor of production is culture that generates new business through the activation of productive connections among the economic actors. Districts as clusters are obviously studied for different purposes and by means of different methods in respect to districts as projects for local development; therefore, networking measures can be used as a proxy of the degree of consolidation and strengthening of a district.

Tourism flows are also a proxy of level of development as well as of sustainability, but the management of tourist flows is, in many cases, a completely different issue, which concerns the protection of natural and cultural resources by human impacts, eventually too hard or too concentrated in time.

The concept of value chain of cultural heritage is sometimes used to identify actions creating a stronger integration between the production processes of different firms and economic activities, paying attention to the constraint of ensuring the necessary economies of scale to ancillary industries and a demand for their products.

Although the enhancement of cultural resources presents enormous potential for local development, some areas are struggling to obtain significant results, in spite of substantial investment and it is necessary to focus further on the other elements useful to understand the reason of it. Regional development policies may implement measures that are mainly focused on enhancing the attractiveness of local culture for tourism. A limited effectiveness can be caused not only by the management of cultural resources, but also by the difficulty in optimising the other elements of the tourism product provided by the destination (accessibility services, accommodation, catering), their quality levels or, still, the aspects related to their communication. Conversely, all the components and cultural attractions of the area should be integrated within a distinctive image of the tourist destination, properly passed through traditional and innovative marketing channels.

In addition, although cultural tourism is considered a phenomenon that will grow strongly in coming years, tourists are always more demanding and paying attention to the continuous renewal of cultural activities and events through which destinations seek to sustain competitiveness. They are also sensitive to the sustainable management of territories and of natural environment, as well as to authenticity and creativity in tourism experiences (Richards & Wilson, 2006). In this view, the availability of cultural resources is not the only determinant for the success of the destination; the originality of the mixture of cultural resources, the ability to continually renew the cultural program, and appropriate targeting and communication give destination a lasting competitive advantage. It is, in short, the ability to link cultural heritage to cultural industries that allows usable and marketable production, new wealth and job creation. New business initiatives can start-up, based on a creative use of culture and heritage embedded in the historical, artistic and human resources.

Furthermore, culture is a cross-functional input to all productive sectors - like research and information technology - and the pattern of penetration is not predictable a priori. Therefore, the more cultural marketing actions target the residents rather than tourists, the more they represent a long-term investment rather than a quick return promotion of the territory. Nevertheless, the conditions for sustainable development accrue, depending on collective learning, inter-generational transfer of skills and generation of new businesses as innovative cultural experiences.

The high levels of uncertainty that firms producing cultural products typically face in final markets accentuate the network or transactions-intensive character of production, as uncertainty tends to induce high levels of vertical disintegration as a way of reducing intrafirm misallocation of resources (Scott, 2006). Anywhere, there is "little or no room in the analysis … for claims that advanced forms of creativity in cities can be induced simply by making them attractive on the consumption side for individuals with high levels of educational attainment and "talent". Such individuals are incontestably necessary for the effective functioning of creative cities in the modern era, but they by no means represent a full set of sufficient conditions as well. Creativity and its specific forms of expression in any given city are induced in complex socio-spatial relationships constituting the local creative field, which in turn is centrally rooted in the production, employment, and local labour market dynamics" (Scott, 2010).

Building cultural entrepreneurship has the advantage of captured local markets, but it must also be outward-looking, both regionally and globally (UNCTAD 2004). Specific attention has to be paid to the identification and involvement of key stakeholders; although legitimate stakeholders having the right to be involved in a collaboration "must also have the resources and skills (capacity) needed to participate" (Jamal & Getz, 1995, p. 194), the enrolment process should be broad-based, to include all possible categories of stakeholders, but mediated through institutional representatives (e.g. trade and industry associations, mayors of local municipalities, cultural bodies) and fact-building, showing the potential benefits for each stakeholder as well as threats and weakness discouraging unrealistic expectations (Arnaboldi & Spiller, 2011).

CONCLUSION

Cultural district emerges as an innovative model for local development, with a precise conceptual and practical identity, well distinct from other similar forms of territorial clusters which share the same label of "district", such as industrial or tourist districts.

This work highlights why the peculiarities of this model – culture as a catalyst of local resources, territory as a source of creativity, a mission focused on increase of values and cultural assets, a hybrid governance which combines centralized and spontaneous coordination – are distinctive elements that can become key factors in supporting a long term process of sustainable development. In particular, it may be stressed the central role played by culture as the trigger of a virtuous circle which can produce creativity-based innovation and "spread value" for all the stakeholders.

This emphasis on culture overshadows other important purposes in local development, such as the commercial success of local companies or the increase in touristic flows. These are considered as natural outcomes of a successful cultural district, but not as its priorities, as the basic idea is that cultural resources are the heart of the system: a heart that pumps blood throughout the territory, ensuring its survival and growth. This approach requires a long time to express all its potential, but has the indisputable advantage of ensuring the best conditions for a durable and sustainable development, which combines economic and social well-being with environmental protection.

In this sense, the model of cultural district seems to be an ideal solution, particularly fitting for those regions where a rich endowment of cultural resources goes with a lack of business ventures. The problem is that the creation of a cultural district is very awkward, due to the fact that cultural district cannot be "created". They can only emerge as a match between a wise top-down strategy inspired by a long term vision and the bottom-up inventiveness of local stakeholders, which both must found their action on the cultural assets of the territory.

Policymakers should, therefore, avoid the excess of planning which often distinguishes their work and try to take on a less assertive methodology, more respectful of the local community. In this effort they could be inspired by recent theories of strategic management, which have given up the myth of strategic planning, accepting the idea that the formalisation of strategy is a sense-making event that helps to rationalize past decisions, where emergent actions prevail on deliberate intentions (Mintzberg & Waters, 1985). Another good input that policymakers can draw from strategic management theories is the focus on resources and capabilities rather than on abstract plans. Since the 1980s a wide literature on resource-based view (Wernerfelt, 1984; Rumelt, 1984; Barney, 1991) has assumed that the basis for a competitive advantage of a firm lies primarily on the application of the bundle of valuable resources at the firm's disposal, while most recent studies (Teece et al., 1997; Teece, 2007) have pointed out that in a rapidly changing environment the durable success of a firm depends on its dynamic capabilities, that is "the capacity of an organization to purposefully create, extend, or modify its resource base (Helfat et al., 2007). If this is true for firms where hierarchy can ensure a stricter coordination among people, it is much more appropriate for territories where competitive processes are managed by complex networks of independent and heterogeneous organisations.

A strategic plan can never be imposed to a territory as a top-down decision. Policymakers have to understand that a path of sustainable development cannot be the fruit of a wishful desk work that brings into being an abstract design of the future of their territories. They have to carry on a long-term process driven by a strong vision to consolidate and increase the key resources of their region, those which can sustain their competitive advantage, starting from the building of a clear community identity and a solid social capital, the two main components of successful cultural district, together with some distinctive cultural assets. They must create the best conditions to enact the virtuous circle which links together culture, creativity and innovation, as this "circle" will also ensure the development of those dynamic capabilities which can maintain the competitiveness of a territory along time. A region where culture is the lifeblood of the local community will, in fact, be ready to "sense" and "seize" the best opportunities, addressing a continuous "transformation" process in order to keep a strong stock of competitive resources[7]. Creativity and innovation will be a stable attitude within all the district, so that everybody will give a contribution to sustain and renew the overall competitiveness of the system.

The development of effective conceptual models and useful management tools that can help policymakers to interpret the dynamics of the process of generation and functioning of a cultural district may be a challenge for practitioners but also for future research. Actually, if the first aim of policies must be the "facilitation" of these process, the decision-makers cannot face such a task without a deep understanding of the phenomenon, which is not yet provided by present theories.

NOTES

1) Evans (2009) suggests a classification of cultural quarters and creative quarters, defining specific features depending on an economic, a social and a cultural framework. The first type is founded on a process of local economic development with a high range of place-making branding, where the zoning and the regeneration, in terms of "culture", are key elements of orientation. These cultural quarters have a high level of historic preservation and conservation and are identified as festival and cultural centers in cultural city. The second type of creative quarter is mixed-used, with more diversity and urban design quality in terms of buildings, facilities and landscapes. They have an area of polarization and attractiveness expanded on the city-region and they are based on the knowledge economy. They produce new high-technology services, creative products as well as innovation spillovers. Often cultural and creative characteristics are present in the same creative and cultural clusters which develop a multi-dimensional identity and multi-functional uses.

2) The explicit reference to the concept of "district" within tourism industry has been introduced by Santarelli (1995), who used it to describe the specific situation of the Adriatic Coast of Romagna and Marche. Antonioli Corigliano (1999) applied the concept to food and wine

tourism, where the boundary between tourism and manufacturing activities appears, moreover, particularly ambiguous. In 2001, then, ACI-Censis study has provided a systematic mapping of the Italian "tourist districts" (ACI-Censis, 2001). The term district is present in Anglo-Saxon literature (Stansfield & Rickert, 1970; Judd, 1993; Pearce, 1998), but generally refers to a neighborhood of a metropolitan area (*urban district*) in a different meaning from that used in most of Italian literature, where the regional scope is much broader and refers to the concept of a tourist destination. Other similar territorial models of systemic approaches applied to the tourist industry are that of *tourist milieu* (Michalkó & Rátz, 2008), deriving from the French and Swiss literature, or that of *tourist cluster* (Gordon, Goodall, 2000; Van Den Berg et al., 2001; Svensson et al., 2006), inspired by Porter's work (1998).

3) Even Marshall in his first conceptualization stresses the importance of time for the spontaneous development of a district, a place where "the mysteries of the trade become no mysteries; but are as it were in the air, and children learn many of them unconsciously"(Marshall, 1920, p. 271). Only time may contribute to the birth of a district, while planning intentions cannot play a key role.

4) Literature on touristic districts is strictly connected to the mainstream of Destination Management (DM), flourished in the 1990s (Ritchie, 1993; Buhalis, 2000).

5) Even this distinction is characterized by a large ambiguity: in fact, numerous studies, especially abroad, have given rise to a variety of models and estimation procedures, which do not always agree on the definition of the types of expenditure. For instance, it has been pointed out how it should be considered separately in the analysis the expenditure made by visitors from that sustained by the organisers (IReR, 2006, pag. 35).

6) "Culture-based creativity is an essential feature of a post-industrial economy. A firm needs more than an efficient manufacturing process, cost-control and a good technological base to remain competitive. It also requires a strong brand, motivated staff and a management that respects creativity and understands its process. It also needs the development of products and services that meet citizens' expectations or that create these expectations. Culture-based creativity can be very helpful in this respect" (KEA, 2009, p. 5).

7) In Teece's theoretical model, sensing, seizing and transforming are actually the three fundaments of dynamic capabilities, which can allow a firm to develop and consolidate its resource base (Teece, 2007). Of course, this approach may be extended to the territorial strategy too.

REFERENCES

1. ACI-Censis (2001). *Rapporto Turismo. I distretti turistici italiani: l'opportunità di innovare l'offerta.* Censis Servizi, Retrived from: http://www.aci.it/fileadmin/documenti/studi_e_ricerche/monografie_ric erche/ Rapporto_turismo_2001_sintesi.pdf

2. Antonioli Corigliano, M. (1999). *Strade del vino ed enoturismo. Distretti turistici e vie di comunicazione*, Franco Angeli, ISBN 978-884-6413-21-5, Milano, Italy.

3. Arnaboldi, M. & Spiller, N. (2011). Actor-network theory and stakeholder collaboration: The case of Cultural Districts, *Tourism Management,* Vol. 32 (2011) 641-654, ISSN: 0261-5177

4. Barney, J. B. (1991). Firm resources and sustained competitive advantage. *Journal of Management*, Vol. 17, No. 1, (March 1991), pp. 99 –120, ISSN 0149-2063

5. Battaglia A. & Tremblay D.G.(2011) *El Raval And Mile End: A Comparative Study of Two Cultural Quarters in Barcelona and Montreal, Between Urban Regeneration and Creative Clusters.* Research note n. 1°, Canada Research Chair on the Socio-Organizational Challenges of the Knowledge economy, Télé-université/Université du Québec à Montréal, March

6. Becattini, G. (1987). *Mercato e forze locali: il distretto industriale*, Il Mulino, 978-881-5014-20-7, Bologna, Italy.

7. Becattini, G. (Ed.) (1989). *Modelli locali di sviluppo*, Il Mulino, ISBN 978-881-5020-64-2, Bologna, Italy

8. Bellandi, M. (1982). Il distretto industriale in A. Marshall. *L'Industria*, Vol. 3, pp. 355-375, ISSN 0019-7416

9. Belussi, F. & Caldari, K. (2009). At the origin of the industrial district: Alfred Marshall and the Cambridge school. *Cambridge Journal of Economics*, Vol. 33, Issue 2 (March 2009), pp. 335-355, ISSN 0309-166X

10. Bramwell, B., & Sharman, A. (1999). Collaboration in local tourism policymaking, *Annals of Tourism Research*, 26, 392-415, ISSN: 0160-7383

11. Brusco, S. (1989). *Piccole imprese e distretti industriali: una raccolta di saggi*, Rosenberg & Sellier, ISBN 978-887-0113-39-6, Torino, Italy.

12. Bucci, A. & Segre, G., (2009). *Human and Cultural Capital Complementarities and Externalities in Economic Growth.* Paper presented at the conference *Arts, Culture and the Public Sphere*, Venice, 4 - 8 November 2008. Working Paper n. 2009-05 Dipartimento di Scienze Economiche Aziendali e Statistiche, Università degli Studi di Milano

13. Buhalis, D. (2000). Marketing the competitive destination of the

future. *Tourism Management*, Vol. 21, No. 1, (February 2000), pp. 97-116, ISSN 0261-5177

14.   Carta, M. (2002). Strategie per lo sviluppo regionale: l'armatura dei sistemi culturali locali in Sicilia. *Urbanistica Informazioni*, n.185, ISSN: 0392-5005

15.   Carta, M. (2004). Strutture territoriali e strategie culturali per lo sviluppo locale. *Economia della Cultura*, a. XIV, n.1, pp. 39-55, ISSN: 1122-7885

16.   Ciegis, R.; Ramanauskiene, J. & Martinkus, B. (2009). The Concept of Sustainable Development and its Use for Sustainability Scenarios. The Economic Conditions of Enterprise Functioning, *Inzinerine Ekonomika-Engineering Economics*(2). ISSN 1392-2785

17.   Commission of the European Communities (2007) *European agenda for culture in a globalizing world*, Communication 242, Brussels, 10.5.2007, SEC(2007) 570 Retrieved from http://eurlex.europa.eu/LexUriServ/LexUriServ.do?uri=COM:2007:024 2:FIN:EN: PDF on June, 12th, 2008.

18.   Commission of the European Communities (2007). *Agenda for a Sustainable and Competitive European Tourism*. Communication 621,. Retrieved on December, 20th, 2007, from http://eurlex.europa.eu/LexUriServ/LexUriServ.do?uri=COM:2007:062 1:FIN:EN: PDF

19.   Council of The European Union (2006). *Renewed Eu Sustainable Development Strategy*, Brussels, 26 June 2006, 10917/06

20.   Dei Ottati, G. (1986). Distretto industriale, probl 0391-2078emi delle transazioni e mercato comunitario: prime considerazioni, *Economia e Politica Industriale*, Vol. 51, pp. 93- 121, ISSN 0391-2078

21.   Evans, G. L. (2009). Creative Cities, Creative Spaces and Urban Policy. *Urban Studies* 46 (5/6), pp. 1003-1040, ISSN: 0042-0980

22.   Florida, R. (2002). *The Creative Class, The Rise of the Creative Class. And How It's Transforming Work, Leisure and Everyday Life*, Basic Books, ISBN 978-046-5024-76-6, New York, USA

23.   Gordon, I. & Goodall, B. (2000). Localities and Tourism. *Tourism Geographies*, Vol. 2, No. 3, pp. 290-311, ISSN 1461-6688

24.   Helfat, C., Finkelstein, S., Mitchell, W., Peteraf, M., Singh, H., Teece, D. & Winter, S. (2007). *Dynamic Capabilities: Understanding Strategic Change in Organisations*, Blackwell Publishing, Malden, USA, ISBN 978-140-5135-75-7

25.   Helm, D. (1998). The assessment: environmental policy – objectives, instruments and institutions. *Oxford review policy*, 14(4), ISSN 0266-903X

26.   IReR - Istituto Regionale di Ricerca della Lombardia (2006). *Metodologie di valutazione di impatto degli interventi culturali II fase. Rapporto Finale* (Codice IReR: 2006B001), Dicembre.

27.   Jacobs, M. (1995). Sustainable Development – From Broad Rhetoric to local Reality. *Conference Proceedings from Agenda 21 in Cheshire*, 1 December 1994, Chesire County Council, Document No. 49.

28.   Jamal, B. T. & Getz, D. (1995). Collaboration theory and community tourism planning. *Annals of Tourism Research*, 22(1), 186-204, ISSN: 0160-7383

29.   Judd, D.R. (1993). Promoting Tourism in US Cities. *Tourism Management*, Vol. 16, No. 3, pp. 175–187, ISSN 0261-5177

30.   Kates, R.W., Parris, T.M. & Leiserowitz, A.A. (2005). What is Sustainable Development? Goals, Indicators, values, and practices. *Environment: Science and Policy for Sustainable Development*, Vol. 27, N. 3, Pag 8-21. 2005

31.   KEA European Affairs (2006). *The Economy of Culture in Europe. Study for the European Commission*, Directorate General for Education and Culture, Bruxelles, October.

32.   KEA European Affairs (2009). *The Impact Of Culture On Creativity*, A Study prepared for the European Commission (Directorate-General for Education and Culture), June 2009. Retrived from: http://ec.europa.eu/culture/keydocuments/doc/study_impact_cult_creati vity_06_09.pdf on

33.   Landry, C. (2000). *The Creative City: A Toolkit for Urban Innovators*, London, Earthscan, ISBN: 1853836133 9781853836138

34.   Lazerson, M.H. & Lorenzoni, G. (1999). The firms that feed industrial districts: A return to the Italian source. *Industrial and Corporate Change*, Vol. 8 No. 2, (June 1999), pp. 235- 266, ISSN 0960-6491

35.   Lazzeretti, L. (2001). I processi di distrettualizzazione culturale della città d'arte: il cluster del restauro artistico a Firenze. *Sviluppo Locale*, VIII, 18, pp. 61-85, ISSN 1974-2193

36.   Marshall, A. (1920). *Principles of Economics*. Macmillan, London, England

37.   Michalkó G., & Rátz, T. (2008). The role of the tourist milieu in the social construction of the tourist experience. *Journal of Hospitality Application and Research*, Vol. 3, No.1, pp. 22- 32, ISSN 0973-4538

38.   Mintzberg, H &. Waters, J.A. (1985). Of strategies, deliberate and emergent. *Strategic Management Journal*, Vol. 6, No. 3, (July-September 1985), p. 257-272, ISSN 0143- 2095

39. Normann, R. (1984). *Service management: strategy and leadership in service business*, Wiley, ISBN 978-047-1904-03-8, Chichester, UK.

40. Pearce D.G. (1998). Tourist districts in Paris: Structure and functions. *Tourism Management,* vol. 19, No. 1, pp. 49–65, ISSN 0261-5177

41. Piore, M. J., & Sabel, C. F. (1984). *The second industrial divide: Possibilities for prosperity.* Basic Books, ISBN 978-046-5075-61-4, New York, USA

42. Porter, M. (1998). *The Competitive Advantage of Nations.* Free Press, ISBN 978-068-4841-47-2, New York, USA

43. Richards, G. & Wilson, J. (2006). Developing creativity in tourist experiences: A solution to the serial reproduction of culture? *Tourism Management,* 27, pp. 1408–1413, ISSN: 0261-5177

44. Ritchie, J.R.B. (1993). Crafting a Destination Vision: Putting the Concept of Resident- Responsive Tourism into Practice. *Tourism Management,* Vol. 14, No. 5, pp. 379-389, ISSN 0261-5177

45. Roodhouse, S. (2006). *Cultural Quarters. Principles and Practice.* Bristol, Intellect Books, ISBN 9781841501581

46. Rumelt, R.P. (1984), Rumelt, R. P. (1984). Towards a strategic theory of the firm. In: *Competitive strategic management*, Lamb, B. (Ed.), pp. 556–570, Prentice-Hall, ISBN 978-013-1549-72-2, Englewood Cliffs, NJ, USA.

47. Sacchetti, S. & Tomlinson, Ph.R. (2009). Economic Governance and the Evolution of Industrial Districts Under Globalization: The Case of Two Mature European Industrial Districts. *European Planning Studies,* Vol. 17 Issue 12, (December 2009), pp. 1837-1859, ISSN 0965-4313

48. Sacco, P. & Ferilli, G., (2006). *Il distretto culturale evoluto nell'economia post industriale.* Università Iuav di Venezia, DADI Dipartimento delle Arti e del Disegno Industriale, WP. N. 4/06, luglio.

49. Sacco, P. L. & Pedrini, S. (2003). Il distretto culturale: mito o opportunità?. Dipartimento di Economia "S. Cognetti de Martiis", International Centre for Research on the Economics of Culture, Istitutions, and Creativity, Università di Torino, *Working Paper Series,* n. 05.

50. Sacco, P.L. & Segre. G. (2009). Creativity, Cultural Investment and Local Development: A New Theoretical Framework for Endogenous Growth. U. Fratesi and L. Senn (eds.) *Growth and Innovation of Competitive Regions – The Role of Internal and External Connections,* Springer-Verlag Berlin Heidelberg 2009, pag. 281- 294, ISBN 978-3-642 08991-6

51. Santagata, W. (2002). Cultural districts, property rights and sustainable economic growth. *International Journal of Urban and Regional Research*, 26/1, pp. 9-23, ISSN: 1468-2427

52. Santagata, W. (2005). I Distretti culturali nei Paesi Avanzati e nelle Economie Emergenti. *Economia della Cultura*, a. XV, n. 2, pp 141-152, , ISSN: 1122-7885

53. Santarelli, E. (1995). Sopravvivenza e crescita delle nuove imprese nei distretti industriali. Il settore turistico nel Medio Adriatico. *L'Industria*, Vol. 16, No. 2, pp. 349-362, ISSN 0019-7416

54. Scott , A. J. (2006). Entrepreneurship, innovation and industrial development: geography and the creative field revisited. *Small Business Economics* 26 (1): 1–24, ISSN 0921- 898X

55. Scott , A. J. (2010). Cultural economy and the creative field of the city. *Geografiska Annaler: Series B, Human Geography* 92 (2): 115–130, ISSN: 1468-0467

56. Stansfield, C. & Rickert, J. (1970). The Recreation Business District. *Journal of Leisure Research,* Vol. 2, No. 4, pp. 213–235, ISSN 0022-2216

57. Stern, M. J. & Seifert, S.C. (2007). Cultivating "Natural" Cultural Districts, *Creativity & Change,* September, 2007, Social Impact of the Arts Project, University of Pennsylvania's School of Social Policy & Practice, Retrieved on September, 11th 2011, from http://www.trfund.com/resource/downloads/creativity/NaturalCulturalDi strict s.pdf

58. Svensson B.; Nordin, S & Flagestad A. (2006). A governance perspective on destination development-exploring partnerships, clusters and innovation systems. *Tourism Review*, Vol. 60, No. 2, pp. 32 – 37, ISSN: 1660-5373

59. Teece, D. J.; Pisano, G. & Shuen, A. (1997). Dynamic capabilities and strategic management. *Strategic Management Journal,* Vol. 18, No. 7, pp. 509-533, ISSN 0143-2095

60. Throsby D., (2004). Assessing the Impacts of the Cultural Industry. Conference Proceedings *Lasting Effects: Assessing the Future of Economic Impact Analysis of the Arts,* May 12- 14th, Tarrytown, NY, Cultural Policy Center, University of Chicago, Retrieved on September, 1st, 2010 fro http://culturalpolicy.uchicago. edu/papers/2004-lasting-effects/Throsby2.pdf Throsby, D. (2001). *Economics and culture.* Cambridge University Press, Cambridge, UK,ISBN-10: 0521586399

61. Tremblay, D.G.; Klein, J.L. & Fontan, J.M. (2009). *Initiatives Locales et Développement Socioterritorial.*, Télé-université, Université du Québec à Montréal, Québec, ISBN 978-2-7624-2250-4

62.   UNCTAD United Nations Conference on Trade and Development (2004). *Creative Industries and Development*. Eleventh session, São Paulo, 13–18 June 2004 Distr. General TD(XI)/BP/13 pag. 1-13

63.   United Nations World Commission on Environment and Development – WCED. (1987). *Our Common Future* (United Nations Report). Oxford: Oxford University Press.

64.   Valentino, P. A. (1999). Strategie innovative per uno sviluppo economico locale fondato sui beni culturali. Valentino P. A., Musacchio A., Perego F., (a cura di), *La Storia al Futuro. Beni Culturali, specializzazione del territorio e nuova occupazione*, Giunti, Firenze, ISBN8809014286

65.   Van Den Berg, L.; Braum, E. & Van Winden W. (2001). Growth Clusters in European Cities: An Integral Approach. *Urban Studies*, Vol. 38, No. 1, (January 2001), pp. 185-205, ISSN 0042-098

66.   Varaldo, R. (2004). Competitività, economie locali e mercati globali: alle radici del declino industriale e delle vie per contrastarlo. *Economia e politica industriale*, Vol. 121, pp. 43-65, ISSN 0391-2078

67.   Wernerfelt, B. (1984). A Resource-Based View of the Firm. *Strategic Management Journal*, Vol. 5, No. 2. (April- June, 1984), pp. 171-180, ISSN 0143-2095

68.   World Summit on Sustainable Development (2002). *From our origins to the future* , Declaration on Sustainable Development, Johannesburg, South Africa, 2- 4 September 2002,
http://www.un.org/esa/sustdev/documents/WSSD_POI_PD/English/P OI_PD. htm

69.   Yusuf, S. & Nabeshima, K. (2005). Creative Industries in East Asia. *Cities*, Vol. 22, N. 2, PP. 109-122, ISSN: 0264-2751

CHAPTER 5

Preservation and Conservation of Rural Buildings as a Subject of Cultural Tourism: A Review Concerning the Application of New Technologies and Methodologies

Cano M1 *, Garzón E[1] and Sánchez-Soto PJ[2]

[1] Department of Rural Engineering, University of Almeria, La Canada de San Urbano, Spain

[2] Instituto de Ciencia de Materiales, Centro Mixto CSIC–US, Spain

ABSTRACT

The cataloguing and promotion of rural architecture contribute to creating jobs by stimulating new economicactivity, such as the promotion of cultural tourism, while preserving a valuable source of information on rural culture, recovering local construction techniques, encouraging a sense of community, and making villages and rural areas more attractive to visitors.The general aim of the present Review is to analyse agricultural and rural buildings and their properties in order to identify key issues concerning sustainable reutilisation in tourism.

It is intended to Almeria province (Andalucía, Spain) a more complete sustainable tourism development framework to satisfy all the demands, from sun and beach at the coast as well as alternative tourism at the interior, based on the protection and the valuable re-utilization of popular rural architectural heritage with potential interest for tourism purposes. Thus, a catalogue of traditional rural buildings in a particular area was carried out, identifying and characterizing each one, establishing criteria for a dynamic and rational selection. The modelling of the lifecycle for the architectural project carried out, for the reuse of rural heritage, as a way of stimulating the tourism sector, is based on methods and risk analysis (statistical methods) and multicriteria analysis that allow the reduction in the subjectivity of the deterioration models carried out. Through the application of the Analytic Hierarchy Process (AHP), as a weighing method, a linear, weighting and additive model that combines all of the factors in one unique global assessment can be obtained, for which values have been assigned to the coefficients of the linear expression that reflect the relative importance of each factor.

Keywords: Rural tourism; Rural buildings; Cataloguing; GIS

INTRODUCTION

From the sixties of last century, the model of tourism at the Mediterranean coast has been sun and beach, being characterized by the existence of lot of people with acquisition power at medium-low level and specially concentrated in determined areas, in particular during summer times. During last decades, this phenomenon has changed by diverse reasons, for instance by a mentality change and the increase of free time [1].

The expansion of rural tourism is a trend that is common to most countries in Europe. Tourism is considered to be a potentially complementary activity for local communities and especially for farming families. The benefits are generally summed up as a three-way yield for the host community (the economical and social dimension of rural tourism), for the land itself (environmental maintenance), and for the tourist (leisure and tourism in the countryside), which implies a sequence of inter-related benefits. All these elements place this type of tourism within a framework of long-lasting development [1].

Tourism is considered to be one of the diversification tools in rural development [2,3]and sustainability discourses have been expanded to tourism [4]. Sustainable tourism, according to the World Tourism Organization, is: "envisaged as leading to management of all resources in such a way that economic, social and aesthetic needs can be fulfilled while maintaining cultural integrity, essential ecological processes, biological diversity, and life support systems" [5].

Sustainability in a popular building context is determined by several factors. A primary factor is that the construction industry is responsible for a large proportion of pollutants and material and energy use world-wide [6]. In addition, the qualitative characteristics of actual buildings create a long-term dependency on e.g. how energy-effectively these can be used or how often maintenance/renovation is required. Furthermore, as a result of the embodied energy in construction [7], it is more sustainable to keep and renovate e.g. former popular building rather than pulling them down and erecting new buildings. Alternatively, in the case of e.g. partial demolition, building materials can be efficiently recovered, refitted and reused, thereby reducing the environmental impact [8].

The only chance abandoned, redundant popular buildings have for survival and conservation is through utilization [9].

Monuments are undeniable documents of world history. Their thorough study is an obligation of our era to mankind's past and future. Over the recent decades, international bodies and agencies have passed resolutions concerning the obligation for protection, conservation and restoration of monuments. The Athens Convention (1931), the Hague Agreement (1954), the Chart of Venice (1964) and the Granada Agreement (1985) are only but a few of these resolutions in which the need for geometric documentation of the monuments is also stressed, as part of their protection, study and conservation [10].

The present authors consider that the re-utilization of popular architectural heritage with tourism purposes is very positive by several reasons, mainly the recovery of old rural and industrial buildings, generally under ruination conditions. Their recovery is very valuable to get new improved buildings where an alternative tourism offer can be presented.

UNESCO (1946) and the Council of Europe have formed specialized organizations for conservation of Cultural Heritage. ICOMOS (International Council for Monuments and Sites) is the most important one, but also CIPA (International Committee for Architectural Photogrammetry), ISPRS (International Society for Photogrammetry & Remote Sensing), ICOM (International Council for Museums), ICCROM (International Centre for the Conservation and Restoration of Monuments) and UIA (International Union of Architects) are all involved in conservation task of cultural heritage [10], andthe IAPH (the Andalusian Institute of Historic Heritage), all they are involved in conservation works concerning Cultural Heritage.

In this sense, García and Ayuga[9], point out some of the most relevant positive consequences of reusing rural buildings in Spain, such as saving energy and materials, creating new jobs and economic activities, promoting cultural tourism, preserving a valuable source of documentation on rural culture, recovering local construction techniques, encouraging a sense of community and improving villages and rural scenery, making them more attractive to visit.

On the other hand, over the last few decades the European countryside has undergone profound changes. The mechanization of the agricultural sector, the modernization of facilities, the demographic shift to the cities and the ageing of the remaining inhabitants of rural areas, have brought about the decline in use and consequent dereliction of a large part of the existing traditional architecture [11]. This phenomenon is common in the rural areas of most European countries

[12]. Nevertheless, we should highlight the social and cultural relevance of traditional buildings. Historic urban sites and traditional houses are the most important evidence of past lifestyles. The conservation of these traditional values within the context of preserving and revitalizing architectural heritage constitutes the preservation of culture [13]. Rural ways of living, until a few decades ago, accounted for a considerable number of land and property owners along with labourers who played a key role in the lifestyle of farming and husbandry. The buildings in these farming population centres are not isolated elements, but fit within a territorial and socio-cultural context, identifying construction methods, materials, styles, aesthetics, etc. produced by a society through the ages. However, the fact that their relevance transcends what is merely material and tangible means that this legacy can be thus preserved [14].

The primary objective of this research is to define the level of structural and architectural adaptability between man and natural environment, at a low population density scale and with a strategic level of social organization that allow their current physical continuity. The problems explained lie in how to achieve sustainable restoration, through the incorporation of multidisciplinary analysis aspects such as: the ecological, cultural and structural spectrums, in an action plan that ranges from defining the problems of each settlement to the recommendations and protective measures to be adopted for each case. The lack of a clear and precise scientific methodology to identify, select and priorities the indicators for assessing the potential of restoring popular architectural heritage has been identified.

The general aim the present review, was to analyse agricultural and rural, using Almería province (Andalusia, Spain) as an example. Specific goals are:

- To identify key factors concerning construction methods and building materials influencing the sustainable reutilisation of agricultural buildings during the renovation-refurbishment process in rural tourism enterprises.

- To account for territorial differences concerning approaches to sustainable agricultural building reutilisation in rural tourism, and study how the location affects these processes.

Methodologies of Study: Research Efforts

One of the main characteristics of rural architecture is their "utilitarian purpose", to the extent that buildings lacking a function have tended to disappear. For this reason, reuse is a viable option for preserving this Heritage [11]. Along these lines [11,12,15-17]have devised methodologies to catalogue traditional buildings, developing identification files, printed or computerized, in which a description of the surroundings appears, as well as the location, type, materials used and graphic documents. Other research work has focused to analyse the buildings detail [11,18-21] for which graphical and visual documentation is particularly important.

A great deal of research has been focused on analysing the detail of the buildings to be reused, where the graphic and visual documentation is particularly important. The problem of decision making has become a subject of significant interest over the last few years in the field of construction. The proof of this is in the appearance of a variety of decision support models and tools, the majority equipped with a particular mathematical device using different approaches, from statistics to fuzzy mathematics, through ways of simplified quantification. The multi criteria methods (MCDA - Multi Criteria Decision Analysis) can be found in this line of work. A vast quantity of studies have been made along these lines: ORME: a methodology for multi criteria classification of buildings [22]; multi criteria material selection procedures for the items in the design project for a building, based on the Analytical Hierarchy Process method (AHP) [23]; study on the adaptive reuse and sustainability of commercial buildings by means of questionnaires and interviews [24]; decision making connected with the planning , design and construction of a building, from the perspective of the problems that the owners and professionals will face whilst considering its reuse, opting to gather information through interviews about the opinions, actions and experiences of all the parties involved in the decision making process regarding adaptive reuse [25].

Research efforts to develop suitable methods for assessing the reuse alternatives of rural buildings have been remarkable. Thus, authors such as Roulet et al. [22], Ipekoglu [13], Zavdskas and Antucheviciene [8], Wang and Zeng [26], Pérez-Martín et al. [21]have developed multicriteria methodologies to reach a multi-dimensional solution. The study of traditional architecture, should include not only an examination of documentation and reuse potential, but also must be focused on a more in-depth analysis considering intangible aspects involved in the restoration of a building, such as a socio-economic study as pointed out by several authors [27-30], as well as the impact on the quality of the landscape where the building is located. Therefore, authors such as Ruda [31], García et al. [32], Hernández et al. [33], García et al. [34], among others, have made a valuable contribution towards this goal.

García and Ayuga [9] highlight point out some of the most relevant positive consequences of reusing rural buildings in Spain, such as saving energy and materials, creating new jobs and economic activities, promoting cultural tourism, preserving a valuable source of documentation on rural culture, recovering local construction techniques, encouraging a sense of community and improving villages and rural scenery, making them more attractive to visit.

New technologies

Additionally, new technologies are being applied in construction, seeking out suitable tools to catalogue every building, for both their quantitative and their qualitative aspects. Ford et al. [35,36] suggested the application of the GIS to historical and geographical data for the analysis of rural architecture. Another example of this is the work by Hernández et al. [34], who have devised new methodologies to study and assess the visual impact of these engineering projects, based on a GIS, developing a program, which is registered by GISCAD 2.0. On this subject, Martín et al. [37] and Garzón et al. [38], indicated that this GIS used as a tool was able to give results in digital format, which can be used in various aspects, such as photographs, diagrams linked to maps, colours or even animation to make these kind of study easier to learn and understand.

GIS is already proven to be extremely helpful and effective in the field of archaeology. It allows archaeologists and technicians to analyze all the existing data and to look for patterns amongst the different layers of spatial data [39]. The GIS process with the goal to survey historic sites and features read cultural and natural values and analyze landscape and urban conservation problems in order to reach a proper approach to preserve, rehabilitate and manage this historical heritage [40].

The use of GIS has an important role in digitizing the base and land surveying maps used for Conservation Aimed Development Plans, preparing database and query and analysis of buildings. Conservation aimed development plans require more detailed and numerous surveying works in various quality rather than all sorts of planning works, to be carried out. Additional new information can be obtained by collecting, drawing (application of the data), evaluating andanalysing of all spatial and non-spatial data in GIS area. In addition to this, the very complicated data obtained can be controlled by means of the queries which are provided by GIS. Having obtained the updating of the stored information, GIS also ease the applications of conservation aimed planning decisions clearly, according to traditional labour-intensive evaluation methods [41].

On the other hand, creating a Web-based GIS environment friendly for end users can maximize the interest of local authorities to use these technological tools [42]. During last years, they have been developing good practices for the record, documentation and management of the information concerning the Cultural Heritage [43].
The modern perception for the methodology of Geometric Documentation is the combination of topometric, topographic and photogrammetric methods, which constitute the most advanced way for a fully controlled survey of the monument of high accuracy [44].

New advances

Recently, best practices have been developed good practices for the record, documentation and management of the information concerning the Cultural Heritage [43].

Pieraccini et al. [45] briefly review state of the art 3D acquisition and digitizing techniques   applied   to   heritage   including   laser   triangulation, stereophotogrammetry, structured light and time of flight. In recent years, 3D digital modeling of heritage works of art through optical scanners has been demonstrated with exceptional results [46]. Arias et al. [47] have proposed computer methods and close-range photogrammetry that allow preventive detection, measurement and tracking of the temporal evolution of structural problems and assessment of the degree of conservation of the materials employed.

Modern methods have become preferable to conventional methods in architecture in its existing state and in the determination of deformations and preparation of measured drawing projects of historical edifices. Digital and 3D data, rich visual images obtained by digital close-range photogrammetry and orthophoto images of edifices are governed by and shepherded in documentation and future conservation projects. Additionally, these methods provide ease, precision and time-savings in measured drawing projects when compared with conventional methods [20].

The modern conception of the geometric documentation methodology is the combination of topometric, topographic and photogrammetric methods, which constitute the most advanced approach for a fully controlled survey of a monument with high accuracy [44].

The authors have very recently proposed a computerized database based on a Geographic Information System (GIS), with hyperlinks to the website for a Rural Development Association (Almeria province, Andalusia, Spain). Thus, a catalogue of traditional rural buildings in this particular area was compiled, identifying and characterizing each one, establishing criteria for a dynamic and rational selection. The purpose to select this example was to facilitate their management by public organizations or private individuals, for their reuse, restoration or both. The cataloguing and promotion of rural architecture contribute to creating jobs by stimulating new economic activity, such as the promotion of cultural tourism, while preserving a valuable source of information on rural culture, recovering local construction techniques, encouraging a sense of community, and making villages and rural areas more attractive to visitors [43].

The severe lack of promotion and cultural activities to complement what the tourist sector has to offer is hindering the development of tourism in a particular region. Therefore, an objective of that recent study was to set in motion a plan to protect vernacular architecture by devising a computerized database based on GIS with hyperlinks to a website. The Almanzora Valley (Almería, Spain) was selected, working in a joint venture with a Rural Development Association, in which a catalogue of traditional rural buildings in the area could be stored, identifying and characterizing each of them, establishing criteria for a dynamic and rational selection of the buildings that are more suitable for reuse and/or

reconstruction. For this purpose, a methodology has been established [43].
The conservation of cultural heritage depends on its education and dissemination. A recent study by the present authors [43] has established the first steps in the protection of rural architectural heritage in the zone under study. It has been investigated the first lines of approach in the identification and classification of the different building typologies and to inform the owners, potential users and corresponding administrators of the possibilities of these buildings. It is clear that there is a long path to establishing opportune measures in the protection of this architectural heritage. This study has intended to disseminate the possibilities of rehabilitation and/or reuse of these constructions, working in a joint venture with a Rural Development Association of cultural tourism interest in the region and with impact at national and international levels.

Application of a New Methodology to Rural Buildings

The recent study by the present authors [43] has established an methodology to study the popular buildings following these working phases: Phase a) Location of the buildings on a cartographic map; Phase b) Field data collection; Phase c) Inventory of rural buildings; Phase d) Socio-economic study of the landscape; Phase e) Creation of identification files, and finally, Phase f) Centralization of documentation in a computerized database.

Area of study

During the last decades, the tourism has been one of the economic activities, besides the technified agriculture focused to the exportation. These both have provided the larger benefits generated for coast regions of South Spain, such as the Almeria province. On the contrary, at the North of Almeria province, there is an economy based on traditional agriculture, some of cattle and scarce industries of diverse typology. In general, they are places without touristic tradition and depressed economically, but characterized by presence of traditional rural buildings (Figure 1) with great potential for reuses as alternative touristic offer.

As shown in Figure 2, the Area of Alto Almanzora is placed in the centre-west-northern of the province of Almería, located by the Almanzora River Valley, the Sierra Filabres to the South limit, and the Sierra of the Estancias to North limit.

The Valley forms a prolonged depression oriented from West to East, constituting a corridor. The zone under study embraced an extension of 1.599 Km2 which are composed of 27 townships, as follows: Albánchez, Albox, Alcóntar, Arboleas, Armuña de Almanzora, Bacares, Bayarque, Cantoria, Cóbdar, Chercos, Fines, Laroya, Líjar, Lúcar, Macael, Oluladel Río, Oria, Partaloa, Purchena, Serón, Sierro, Somontín, Suflí, Taberno, Tíjola, Urrácal and Zurgena. The ensemble of these townships is joined in three sub-area which boundaries coincide with those of Tíjola, Olula del Río-Macael and Albox.

The GIS model

To construct a computerized database, a link was established between the graphics contributed by digital photography and a database managed by Microsoft Excel with ArcGIS, bearing in mind the following aspects of hyperlinking ArcGISto the database created in Microsoft Excel:

Figure 1. Examples of different rural buildings catalogued the Almanzora Valley (Almería, Spain. (Source: adapted by the author).

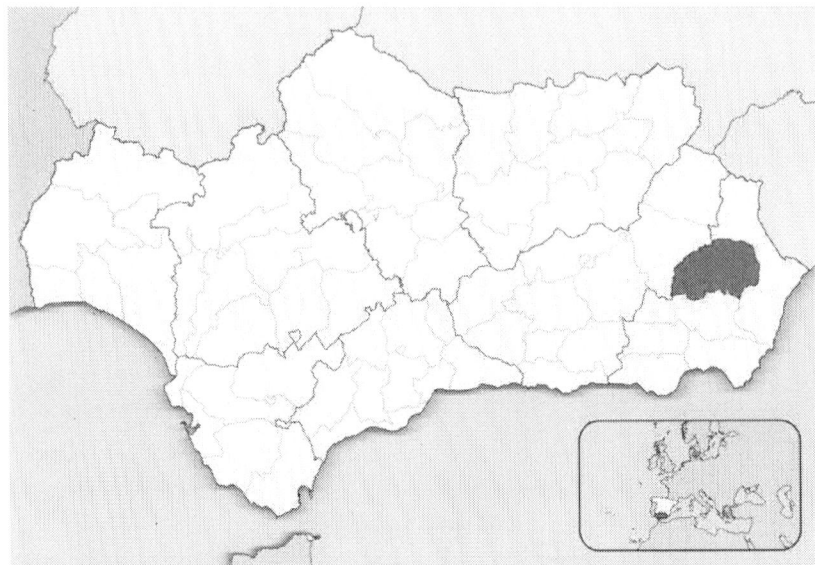

Figure 2. LocalitationtheAlmanzora Valley (Almería, Spain).

1. The application ArcGIS, Arc Map version 9.3, a GIS tool, was used because it has the advantage of linking ArcGIS to photographic documents and the database, and it is very easy to use. It is a unique program which carries out all tasks and avoids incompatibility problems between different applications.

2. The application Microsoft Excel was chosen for the construction of a database. Because it is a program for the creation and administration of databases, we were able to produce a rational database on the buildings in question, which optimized the storage of data and its subsequent retrieval through consultation.

3. To view the photographic documentation, ArcGIS automatically opens each of the photographs using the picture viewer predetermined in the computer. This allows great flexibility when managing photographs, as they can be edited, allocated theme descriptors, retouched, converted to a multimedia presentation, printed, etc.

To devise this tool, all of the digital orthophotography data of the Almanzora region obtained from the National Geographic Information Centre were initially inputted [48]. Once this database was linked, a hyperlink was placed between each building and its identification file, which was also performed in Excel (Figure 3). The photographic documentation of each building was also hyperlinked. The development of the application is shown in Figure 4.

Numerical values in decision-making and guidelines: the GIS-based model

A series of "decision making ratings" have allowed us to study the relationship between the characteristics of the building and its adaptability for reuse, establishing six types of criteria and analyzing 32 indicators (Table 1). In the decision-making process, either an individual or group has to choose between two or more alternatives. These decisions can then lead to solutions of great importance; therefore, it is vital to give them the consideration they deserve.

When problems with decision-making are considered within in a project, in the majority of cases, the problems are multi-criteria. As we are responsible for these problems, the decision making is therefore focused on this approach, making room for Multiple Criteria Decision- Making (MCDM), which considers a range of options (continuous or discrete), different criteria or perspectives and the integration of different factors that are successfully included in the evaluation process. The weight of each criterion is usually previously defined, which is what adds weights to its criticism when speaking about the subjectivity of the approach.

In the MCDM, it is common place for certain aspects to be more relevant to the decision-maker than others. To establish a sound strategy and make correct decisions, it is advisable that the criteria, their relative importance and, in many cases, the indicators are chosen by managers or specialists in their field. Th w1, w2, ..., wney have to define the aspects they consider to be the most important and the general lines of improvement that they have to follow. In this way, it is possible to avoid the decision tree reflecting the position of conflicting parties by placing too much emphasis on certain aspects that are beneficial for some and detrimental for others.

Due to the nature of the present decision-making approach, a finite set of options, the decision based on the diverse nature or attributes of the options with regard to the relevant decision-making criteria, are applicable to multi-objective methods. Amongst these methods, the Analytic Hierarchy Process (AHP) method can be found in the literature [49,50]. The Analytic Hierarchy Process (AHP) method, having a profound impact at a theoretical and practical level, was put forward by Saati in the 1970's. This method comprises the hierarchy tree of decision making, where the first level deals with the main objective of decision making, the second and subsequent ones with the criteria and the final one with the alternatives and possible solutions. For this method to be employed, a great deal of interaction with the main decision-maker is required, in a way that value judgments can be provided at each of the defined hierarchical levels. These judgments involve comparing values in pairs, leading to the provision of $n \cdot (n-1)/2$ value judgments about the relative importance of criteria as well as alternatives. The decision-maker establishes value judgments through Saaty's numerical scale [49] from 1 to 9, comparing in pairs both the criteria and the indicators.

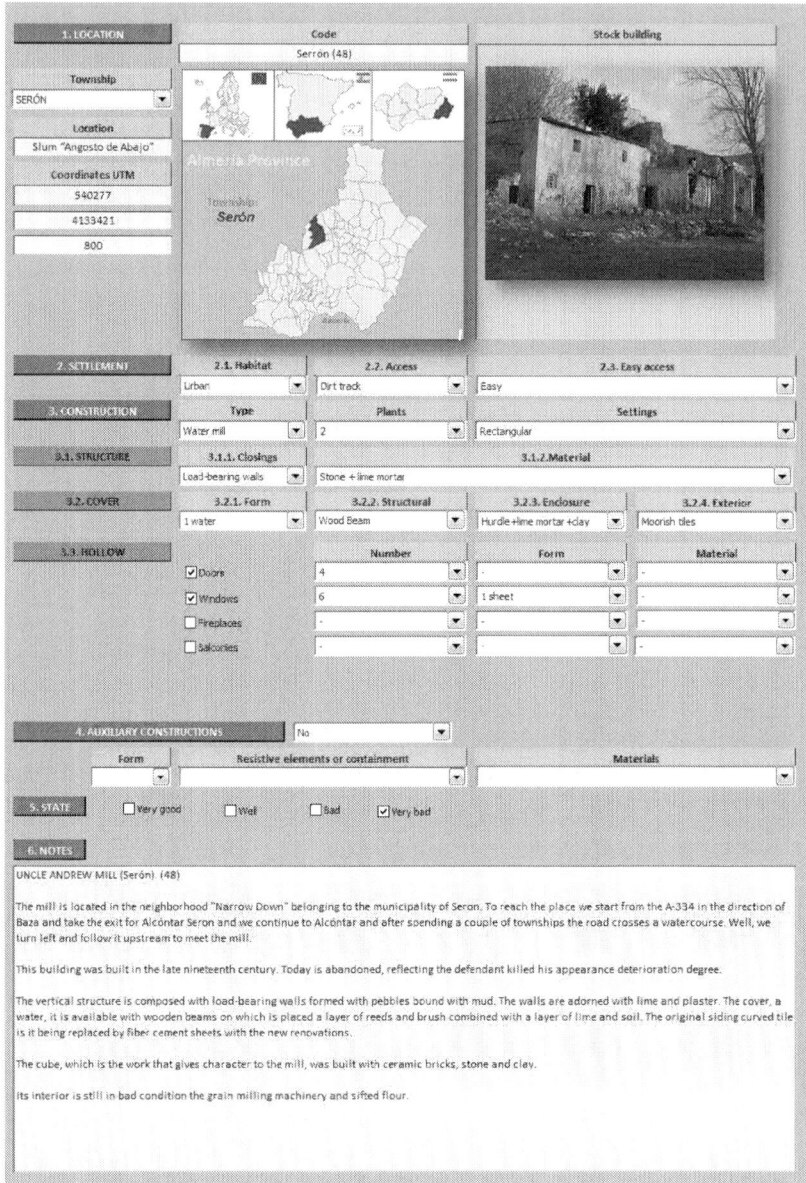

Figure 3. Identification file of a mill. (Source: compiled by author).

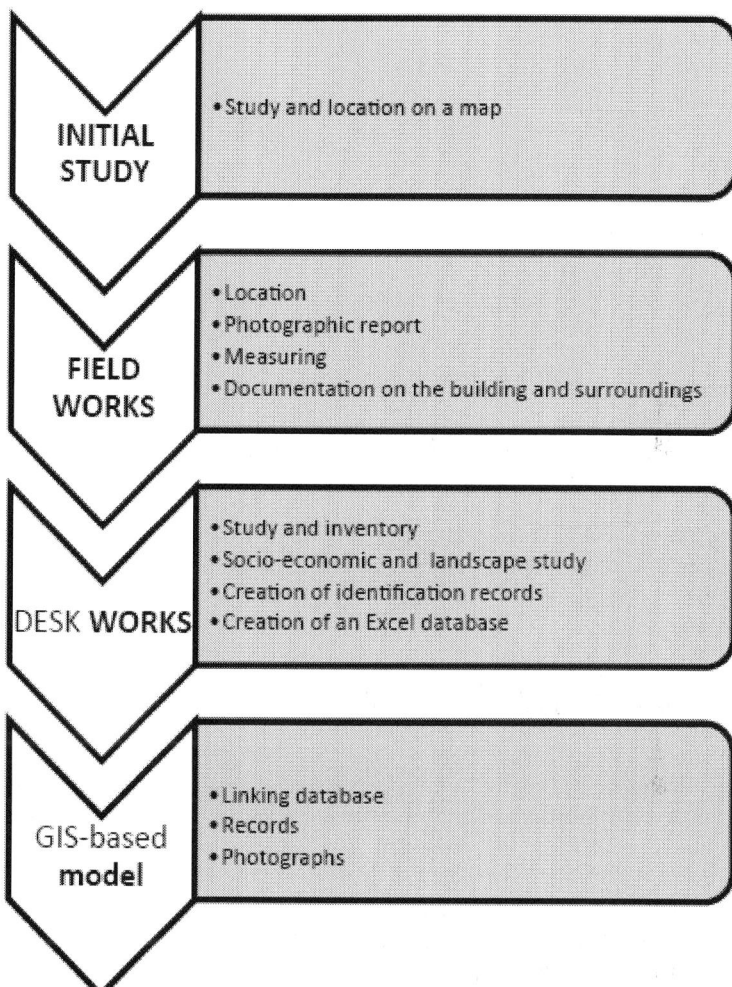

Figure 4. Development diagram of a GIS as a database to catalogue the rural buildings in a district. (Source: compiled by the author).

In this way, a matrix can be built which expresses the relative importance of each indicator selected in rows in relation to the rest of the elements in columns. The decision matrices obtained are squared n x n, n being the number of requirements, criteria or indicators that are being assessed as can be observed in Equation 1. For example, the comparable element in 12 represents the preference for element one over element two. Additionally, it follows that it is a reciprocal matrix as as $ajn = \dfrac{1}{anj}$

Criteria i	C1	C2	.....	Ci	.....	Cn
C1	1	a12	.....	a1i	.....	a1n
C2	1/a21	1	.....	a2i	.....	a2n
....	....	.....	1	.....	.....	.....
Ci	1/ai1	1/ai2	.....	1	.....	ain
....	.....	.....	.....	1	.....
Cn	1/an1	1/an2	.....	1/ani	.....	1

Equation 1. Decision Matrix [A].

From each comparison matrix resulting from each uniform block (requirements, criteria and indicators), the eigenvector for this matrix (Equation 1) defines the weighting for each of the requirements, criteria and indicators used, defined as vector

$$\vec{W} = W_1, W_2, \ldots \ldots W_n$$

The eigenvector is associated with the highest eigenvalue for each decision matrix and represents the ranking or order of priorities. Apart from this, the eigenvalue is a measure of judgment consistency and proves the correct assignment of the preferences. This consistency implies two features: transitivity and proportionality. The first feature shows that the order relations between the elements should be respected if A>C and C>B, so logically A>B. The second feature represents the proportions between the orders of magnitude for these preferences. For example, if A is 3 times greater than C and C is 2 times greater than B, then A must be 6 times greater than B and this would be a 100% consistent judgment.

	Criteria	Indicators
Decision Index	Environment and landscape rating(C_1)	Environment
		Landscape
		Accession
		Topography of theplot
	Surroundings rating(C_2)	Organization of the construction
		Constructive elements of interest
		State of conservation
		Distance to urban core
		Facility of access
	Infrastructure rating(C_3)	Water supply
		Desinfection
		Electrical installation
		State of conservation infrastructures
	Socio-economic rating(C_4)	Tourist attractions
		Services
	General study of the building rating (C_5)	Covered
		Exterior closings
		Interior closings
		Windows/Doors/Soils
		Plumbing/Electricity
		Disinfection
		Singular elements
		Other spaces
	State of the building rating (C_6)	Covered
		Exterior closings
		Interior closings
		Windows/Doors/Soils
		Plumbing/Electricity
		Disinfection
		Singular elements
		Other spaces
		Dampness/Cracks

Table 1. Decision tree with a weight distribution for all variables. (Source: compiled by the authors).

For the application of this method, the different elements need to be structured in a hierarchical way. In the present study, the first hierarchical level corresponds to the Building Decision Index, the second to the Criteria and the third to the Indicators.

If we take the second level of our decision tree and order the six criteria being considered in decreasing order on the basis of their influence on the refurbishment of the construction, it can be presented in the following way:

$$C_3 > C_6 > C_1 > C_4 > C_5 > C_2$$

Next, establish a value function using a measurement scale, such that the numerical system is a scale representation of the relationships between criteria.

The relative weights obtained from the six criteria considered are:

$$\vec{w} = (0.41, 0.20, 0.14, 0, 10, 0.08, 0.07)$$

A weighting consistent with the structure obtained by applying Saaty's AHP method [49]. To obtain the Decision Index of the construction, the indicators pertinent to each criterion are valued. The value of the criterion is subsequently obtained, and finally, the Decision Index of the construction is obtained:

o   The value of the indicators is obtained through quantification according to the values indicated in the field data questionnaires.

o   The value of the criterion is obtained from the value of the indicators pertinent to the same criteria, multiplied by its respective weights, leaving a hierarchical tree like Figure 5.

To obtain the Decision Index of the construction, the indicators pertinent to each criterion are valued.

RESULTS AND DISCUSSION

The GIS-based model developed by the present authors [43] in addition to storing information about the absolute location of the elements (georeferencing) in the system, make possible to calculate distances and areas, maintaining spatial relationships between them, i.e., it puts buildings into groups by type, state, age, location use, etc. This ability is particularly relevant at the resource exploitation phase as it enables us to classify buildings according to attributes. By marking a building on a map, one can see the information about it that is available in the database.

Figure 6 is an example of an identification file and how it is located in digital orthophotography. As shown, the screen is divided into three sections: the top, where the tool bar appears; the left hand window, which contains the layers we have created (villages, mills, traditional buildings, oil presses, livestock trails, paths, tracks, ordinary roads, motorways, riverbeds, rivers, buildings, population, and orthophotographs), which can be activated and deactivated at will; and the main window, where orthophotographs and the activated layers can be viewed. Identification files also appear in the main window, as does the graphic window of each building when we click on its icon in the orthophotograph. Although the identification files of rural buildings developed until now are diverse in terms of criteria and attributes, in the present study, the guidelines established by Fuentes and Cañas [11]have been followed, making up brief files of two pages. These files are easy to complete using drop down menus and are simple to computerize, based on the principle of progressing from general to detailed information. However, at present, it is not possible to prepare general identification files that could be applied to different areas and for different types of buildings, as noted by [11], because extrinsic variables depending on the morphological characteristics of the building and extrinsic factors related to the external context differ enormously between coastal and inland areas. It is possible to standardize the minimum variables that a rural building identification file should consist of. Through the application of the AHP method proposed by Saaty [49]as a weighing method, a linear, weighting and additive model that combines all of the factors in one unique global assessment can be obtained. This is based on the MCDM process described in the methodology section, for which values have been assigned to the coefficients of the linear expression that reflect the relative importance of each factor.

Ultimately, the problem of decision-making in the assessment of potential refurbishment and/or reconstruction of rural buildings lies in creating a suitable system that shows the relationship between the multitude of factors impacting the possibility of refurbishment and, from this, to obtain through parametric adjustment a value function that is a good approximation of the same. Finally, combining all of the partial assessments, a unique mathematical assessment function can thus be obtained that could be supplementary and determine if the variables are independent.

For the evaluation of the refurbishment and/or potential for reconstruction of selected buildings, the present study was applied to 52 samples, after 9 of the initial 61 were rejected as a result of their being in very bad condition or in ruins. The 52 samples principally differed between 4 main types of construction: manor houses, farmhouses, water mills and oil-mills.

In Figure 7, the Decision Index of all of the analyzed constructions is presented, globally depicting the fluctuation of values in the set. The buildings with a lower Decision Index and, therefore, high priority in the reuse order are agro-industrial buildings, mainly oil presses, and to a lesser extent, if they are located near rural enclaves, water mills. As mentioned previously, this result is due to the evolution of productive human activities that tend to be located and developed in more functional and modern buildings. This has led to the abandonment of buildings associated with traditional economic activities, as they have become obsolete because of major changes in production systems, causing a rapid loss of traditional agro-industrial buildings.

It is important to highlight that buildings can be divided into groups by their attributes. The GIS provided selection criterion, for example, "mills with easy access". Then, the layer "mills" can be selected at the top of the window and the code "ACCESS = Easy" is entered. The application will then automatically choose a color, light blue in this case, for all of the buildings that fulfill these conditions (Figure 8). This result demonstrates how useful the GIS are at characterizing and selecting particular buildings based on the specific characteristics of our search.

Another application using the present GIS-based model is the statistical processing of information. If the attribute table of a subject or layer is opened, the attributes to be statistically processed can be selected. For example, we can generate a graph showing traditional buildings according to their state of repair. In this case, as we are dealing with qualitative attributes, it is necessary to provide numeric values to each qualitative criterion established for the state of repair of the buildings. From this statistical processing, we can obtain a report showing the results (Figure 9).

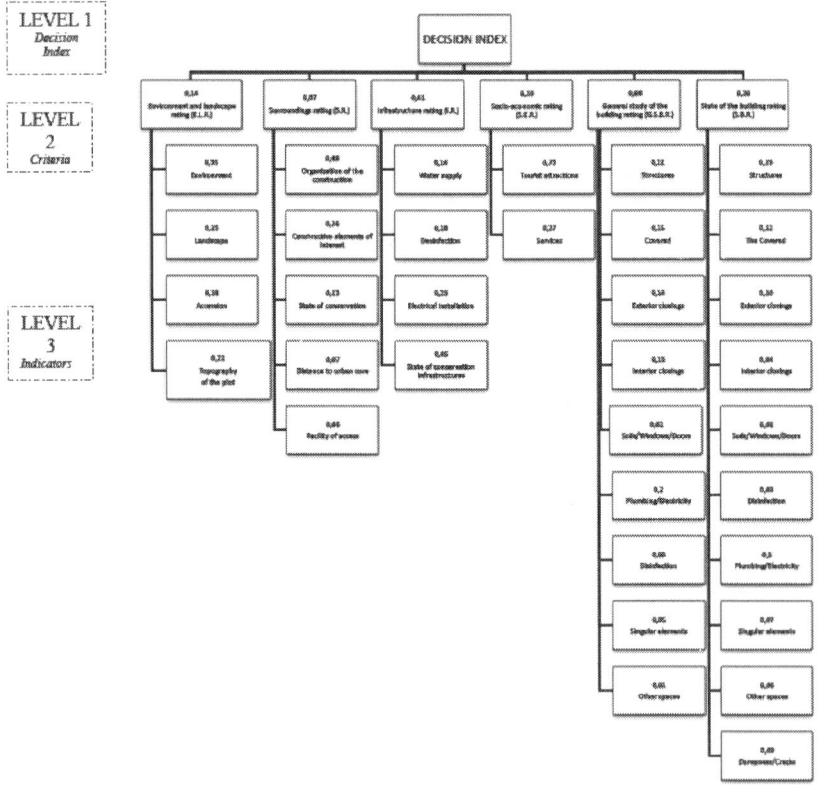

Figure 5. Decision tree with a weight distribution for all variables.(Source: compiled by author).

Finally, the developed GIS-based model enables us to find locations by SATNAV. If we link a sufficiently powerful laptop, PDA or similar tool to a SATNAV aerial, it is possible to establish our location on the screen and devise a possible route for users wishing to visit any of the traditional buildings under study.

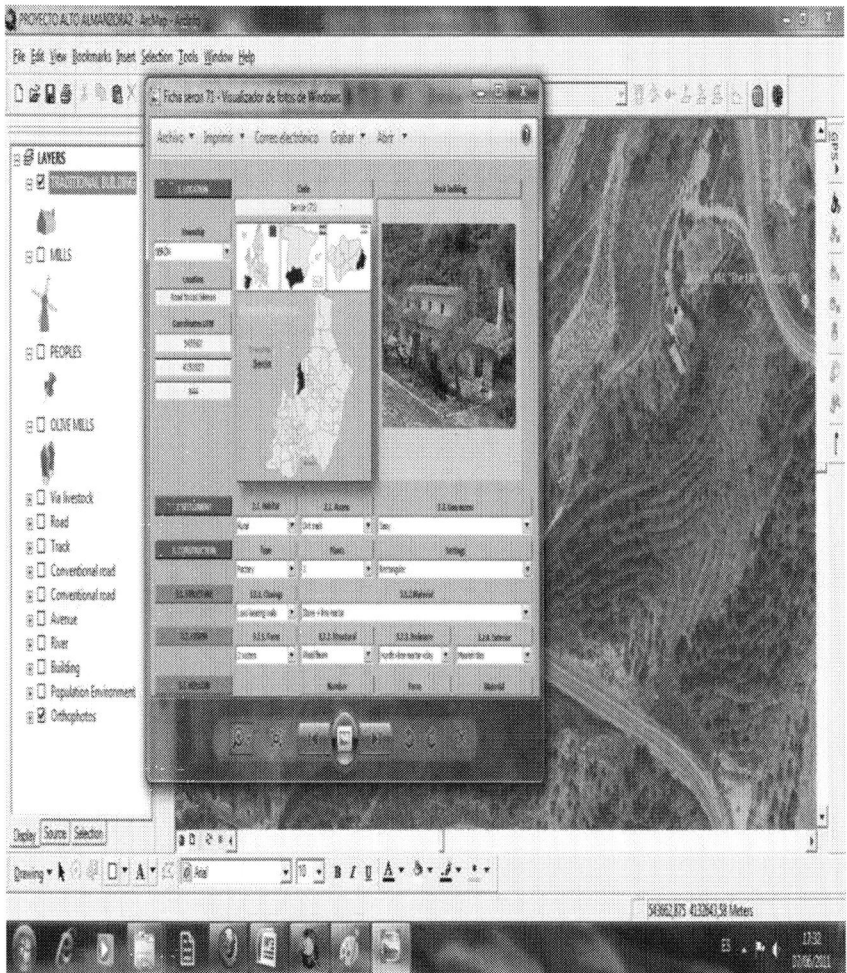

Figure 6. GIS detail, linking the identifying file of a wool factory, after clicking on its location in the orthophotograph of the area. (Source: adapted by the author).

Although authors such as García and Ayuga [9] argue that GIS techniques are not strictly essential to catalogue rural buildings, Fuentes [12], Hernández et al. [33], Ford et al. [35], Ford et al. [36] and Martín et al. [37] suggest using GIS to catalogue historical and geographical data for the analysis of vernacular architecture. Therefore, based on the results obtained, we found GIS to be extremely useful in cataloguing rural architectural heritage. As claimed by Carrera [14], it is clear that we can enter different layers of information that are independent but closely linked to the location of a building, in addition to introducing an architectural approach. This method could also add ethnological, artistic information about places, assets or activities that contain relevant forms of cultural expression and ways of living, providing a tool that helps to coordinate different administrations involved, such as the various ministries, universities and local authorities, thus helping to disseminate valuable results for the tourist trade promotion of this heritage for next generations.

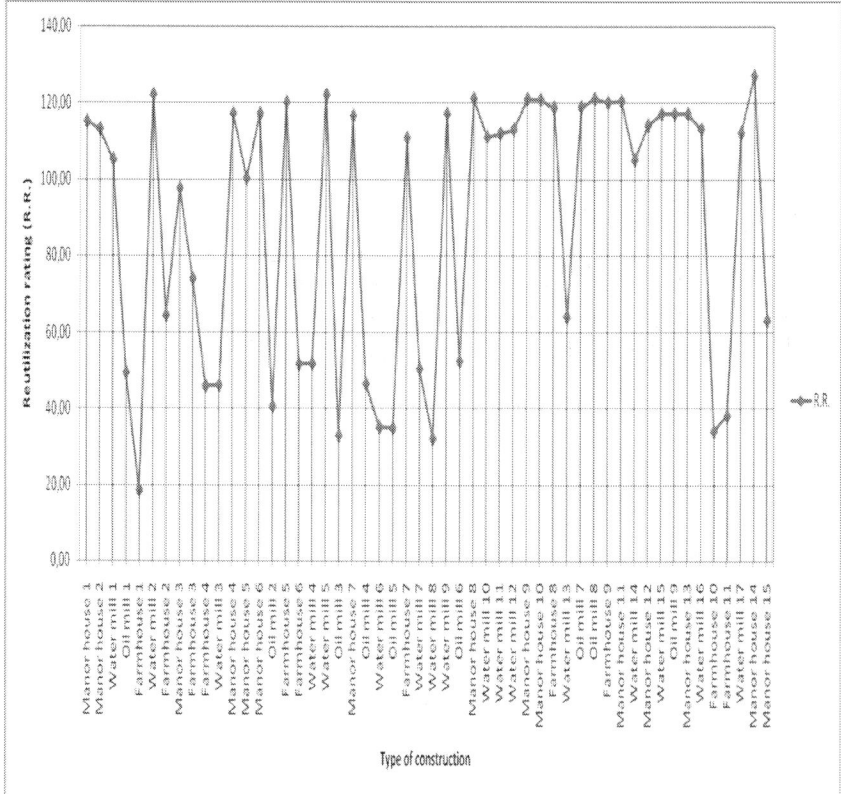

Figure 7. Reutitilization rating of all the analyzed constructions. (Source: compiled by author).

FINAL REMARKS

As shown in the present Review, during last years it has been produced a change in touristic demands. It is originated by a change in mind and the possibility to get more time to spend in idleness. Consequently, the local and regional administrations have tried to increase the touristic offer through the proposal of alternate tourism to that the conventional model of sun and beach.

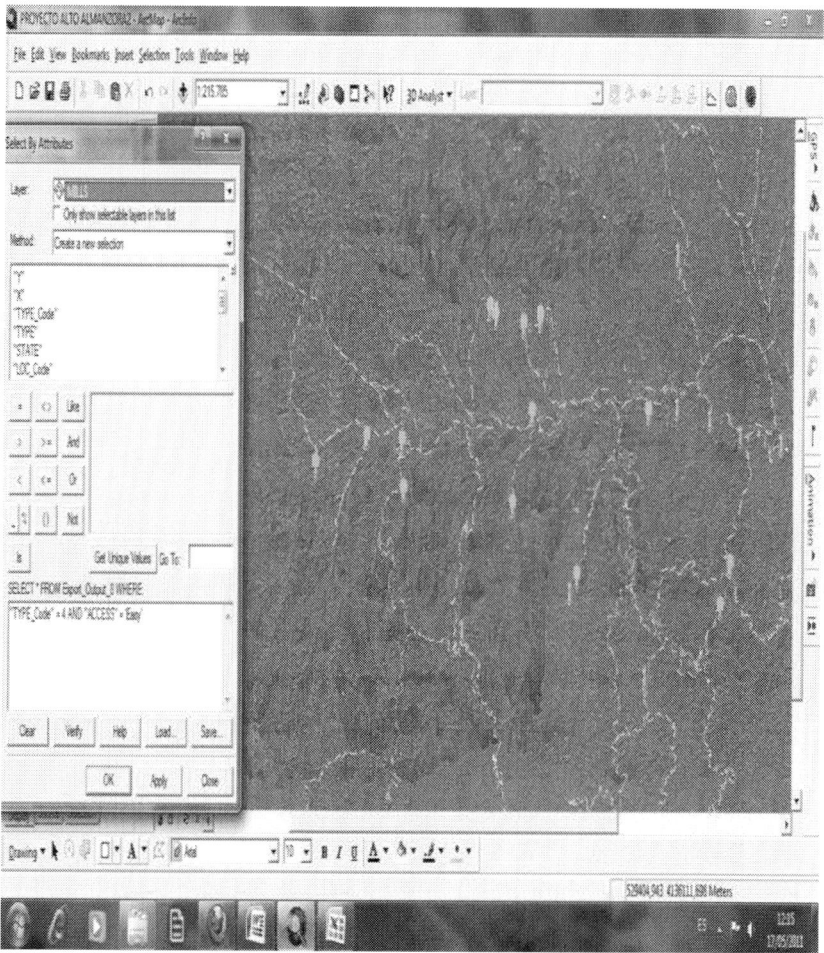

Figure 8. GIS detail, carrying out a search by attributes of the mills, according to the criteria for easy access water mills, which are shown in light blue. (Source: compiled by author).

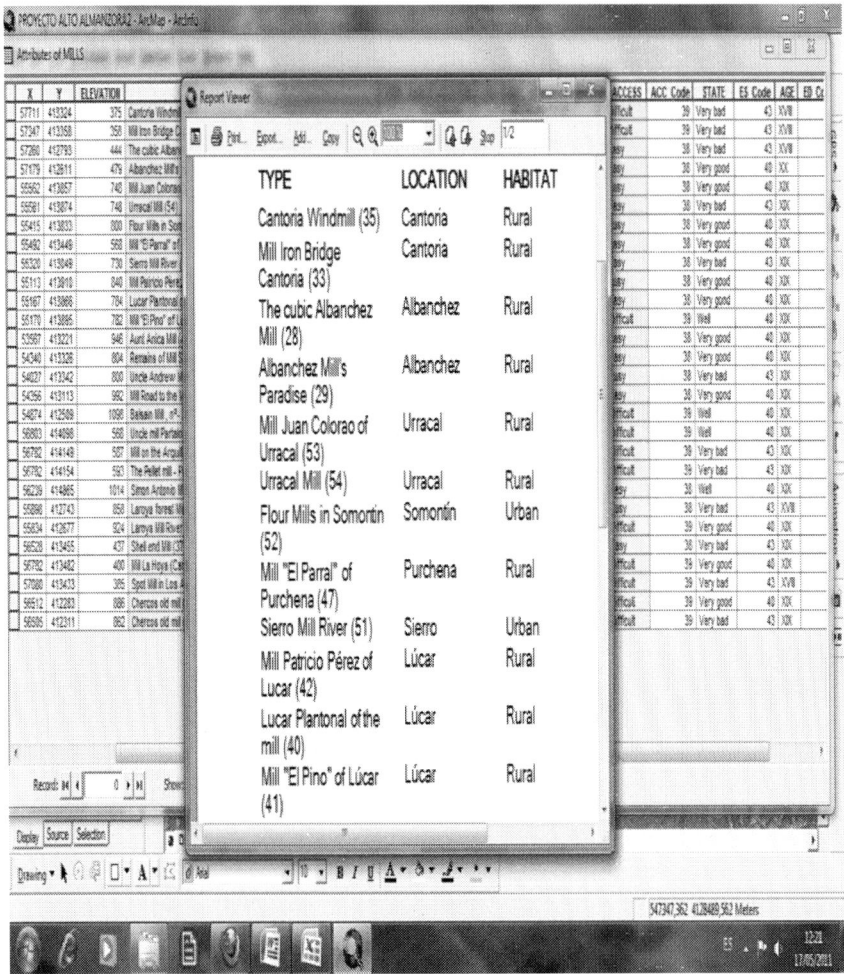

Figure 9. Report on the mills, according to their type, location and surroundings. (Source: compiled by author).

This touristic offer in Almeria province, for instance, has been the way to try the recovery of interior's villages, clearly in economic disadvantage as compared to the coast's villages, which have traditionally been the receivers of tourism. Inside this new mode, all the spaces are potentially considered as tourist trade, although it must be taking into account and to be conscientiously. To avoid any degradation of the environment and states of saturation, it must do a sustainable recreation use of these spaces.

An alternative option of this new offer are several touristic services, such as gastronomy of home-made food prepared using natural local products, generally of own crops; agro-tourism in farms; sport tourism, because the offer is located in natural landscapes which are favourable for several sports: trekking, puenting, canoeing, horseback riding, climbing, etc. In summary, all series of activities which in the traditional model (sun and beach in Almeria province) have not offered to the tourist and during last years, they have claimed.

The proposed methodology provides an efficient method for the inventory and characterization of the buildings in a particular area as an example (Almanzora, Almería province, Spain) based on the GIS. This becomes a working tool which enables us to add new data and to analyse the traditional rural buildings, in such a way that we would not just make the public aware of the region and its potential for tourism, but we would also facilitate the work of devising a possible route for users. Thus, it is to aid for a better development of tourism and promotion of the area, producing a catalogue of rural buildings which will help the management in their reuse or restoration or both by public organizations, or by any private individual interested in doing so. The application Microsoft Excel was chosen as a tool for the construction of a database. Because it is a program for the creation and administration of databases of use every day, it allows greater accessibility for all users in the management of the database.

The refurbishment of traditional buildings in a region entails the appropriate dissemination that can reach everyone. Nowadays, technologies such as the Internet as a tool enable us to make readily available to users, vast quantities of information and data in real time. As a proposal for future research work, it would be a good idea to link the database of the GIS developed elsewhere [43] to a website, as well as being able to download applications such as SATNAV onto a portable device. Doing this, it would be disseminate the catalogue of the traditional buildings of a region, as well as the possibilities of reusing them, especially with reference to their state of repair, refurbishment budgets, quality of the landscape, manpower services and help available for their refurbishment.

All these aims are addressed for a better sustainable tourism development [50].

ACKNOWLEDGEMENTS

The financial support of Andalusian Regional Government (Junta de Andalucía) to Research Groups AGR 107 and TEP 204 is acknowledged. The authors acknowledge the Editor J. Tourism Hospit. for your kind invitation to publish in that Journal.

REFERENCES

1.   Canoves G, Villarino M, Priestley GK, Blanco A (2004) Rural Tourism InSpain: An Analysis of Recent Evolution. Geoforum35: 755-769.

2.   Ilbery B, Bowler I (1998) From Agricultural ProductivismTo Post-Productivism. The Geography of Rural Change.Bowler I. Ed. Harlow, UK: Longman.

3.   Sharpley R, Vass A (2006) Tourism, Farming And Diversification: An Attitudinal Study. Tourism Manage 27: 1040-1052.

4.   OCDE (1994) Environmental Indicators. OECD Core Set. Ed. Paris: OCDE.

5.   World Tourism Organization (2013) Sustainable Development of Tourism.

6.   Bokalders V, Block M (2010) The Whole Building Handbook: How To Design Healthy, Efficient And Sustainable Buildings. London, Earthscan.

7.   Milne G, Reardon C (2011) Embodied Energy.

8.   Zavadskas EK, Antucheviciene J (2007) Multiple Criteria Evaluation of Rural Building's Regeneration Alternatives. Build Environ 42 : 436-451.

9.   García AI, Ayuga F (2007) Reuse of Abandoned Buildings andthe Rural Landscape: The Situation InSpain. T ASABE 50: 1383-1394.

10.   Georgopoulos A, Ioannidis C (2004) Photogrammetric And Surveying Methods For The Geometric Recording of Archaeological Monuments, Archaeological Surveys, FIG Working Week, May 22-27 2004, Athens, Greece.

11.   Fuentes JM, Cañas I (2003) EstudioY Caracterización De La Arquitectura Rural. Obtención, Tratamiento Y Manejo De La InformaciónSobre Las Construcciones. InfConstr 55: 13-21.

12.   Fuentes JM (2010) Methodological Bases for Documenting And Reusing Vernacular Farm Architecture. J Cult Herit11: 119-129.

13.   Ipekoglu B (2006) AnArchitectural Evaluation Method For Conservation of Traditional Dwellings. Build Environ 41 : 386-394.

14.   Carrera G (2004) La ArquitecturaVernácula De La Andalucía Rural. Un AnálisisComparativo De Dos Inventarios.PatrimonioCultural Y Desarrollo Rural En Andalucía 15: 24-29.

15.   Fuentes JM, Gallego E, García AI, Ayuga F (2010) New Uses For Old Traditional Farm Buildings: The Case of The Underground Wine Cellars In Spain. Land Use Policy 27: 738-748.

16.   Sabatino M (2010) Documenting Rural Architecture byGiuseppe Pagano. J Architect Educ63: 92

17.   Martínez A (2004) DesarrolloDe UnaMetodología De Reutilización De ConstruccionesRurales A Partir Del Estudio De Casos. CasoParticular La Provincia De León.Tesis.

18. Armesto J, Gil ML, Cañas I (2006) The Application of New Technologies In Construction: Inventory And Characterisation of Rural Constructions Using The IkonosSatellite Image. Build Environ 41:174-183.

19. Arias P, Ordóñez C, Lorenzo H, Herraez J (2006) Methods For Documenting Historical Agro-Industrial Buildings: A Comparative Study And A Simple Photogrammetric Method. JCultHerit7: 350-354.

20. Yilmaz HM, Yakar M, Gulec SA, Dulgerler ON (2007) Importance of Digital Close-Range Photogrammetry In Documentation Of Cultural Heritage. J Cult Herit8: 428-433.

21. Pérez-Martín E, Herrero-Tejedor TR, Gómez-Elvira Má, Rojas-Sola JI, Conejo-Martin Má (2011) Graphic Study And Geovisualization of The Old Windmills of La Mancha (Spain). ApplGeogr31: 941-949.

22. Roulet CA, Flourentzou F, Labben HH, Santamouris M, Koronaki I, Et Al (2002) ORME: A Multicriteria Rating Methodology For Buildings. Build Environ 37: 579-586.

23. Nassar K, Thabet W, Beliveau Y (2003) A Procedure for Multi-Criteria Selection Of Building Assemblies. Automat Constr12: 543-560.

24. Bullen PA (2007) Adaptive Reuse and Sustainability of Commercial Buildings. Facilities 25: 20-31.

25. Bullen PA, Love PED (2010) TheRhetoric of Adaptive Reuse Or Reality of Demolition: Views From The Field. Cities 27: 215-224.

26. Wang HJ, Zeng ZT (2010) A Multi-Objective Decision-Making Process for Reuse Selection of Historic Buildings.Expert Systems Applications 37: 1241-1249.

27. Bedate A, Herrero LC, Sanz JA (2004) Economic Valuation ofThe Cultural Heritage: Application To Four Case Studies In Spain. JCultherit 5: 101-111.

28. Canoves G, Villarino M, Herrera L (2006) Public Policies, Rural Tourism and Sustainability: A Difficult Balance. BoletinDe La AsociacionDe GeografosEspanoles199-217.

29. Cyrenne P, Fenton R, Warbanski J (2006) Historic Buildings and Rehabilitation Expenditures: A Panel Data Approach. JReal Estate Res28: 349-379.

30. Yiu CY, Leung AYT (2005) A Cost-And-Benefit Evaluation of Housing Rehabilitation. Structural Survey 23: 138-151.

31. Ruda G (1998) Rural Buildings and Environment. Landscape Urban Plan 41: 93-97.

32. García L, Hernández J, Ayuga F (2003) Analysis of the Exterior Colour of Agroindustrial Buildings: A Computer Aided Approach To Landscape Integration. JEnvironManag69: 93-104.

33.   Hernández J, García L, Ayuga F (2004) Integration Methodologies for Visual Impact Assessment of Rural Buildings By Geographic Information Systems. BiosystEng88: 255-263.

34.   García L, Hernández J, Ayuga F (2006) Analysis oftheMaterials And Exterior Texture of Agro-Industrial Buildings: A Photo-Analytical Approach To Landscape Integration. Landscape Urban Plan 74: 110-124.

35.   Ford M, Elkadi H, Watson L (1999) The Relevance of GIS In The Evaluation of Vernacular Architecture. Journal of Architectural Conservation 5: 64-75.

36.   Ford M, Griffiths R, Watson L (2005) The Sand Ford Inventory of Earth Buildings Constructed Using A GIS. Build Environ 40: 964-972

37.   Martín A, Aguilar R, Domínguez V, González JM, Morón ME, Et Al (2005) AplicaciónDe Los SistemasDe InformaciónGeográficaAl Estudio De EdificiosPatrimoniales. CasoPráctico: ActuaciónEn El Humilladero De San OnofreY Su Entorno, San Jerónimo, Sevilla.

38.   Garzón E, García IG, Ruiz-Conde A, Sánchez-Soto PJ (2009) Application of Geographic Information Systems (GIS) In The Search For And Characterization of Raw Materials of Interest In Ceramics And Glass. BoletinDe La Sociedad Española De CeramicaY Vidrio48: 39-44.

39.   Al Bayari O (2005) New Survey Technologies For Production of GIS Model of The Ancient Roman JerashCity.Jordan. Proceedings ofTheCIPA 2005 XX International Symposium, Torino, Italy.

40.   Boriani M, Cazzani A, Giambruno M (2005) The Naviglioof Martesana: A GIS To Manage A Protected Area. Proceedings of TheCIPA 2005 XX International Symposium, Torino, Italy.

41.   Erdem R, Durduran S, Çay T, Dülgerler ON, Yildirim H H(2003) An Experimental Study of GIS - Aided Conservation Development Plan; The Case of Sille- Konya. The International Archives ofThePhotogrammetry, Remote Sensing And Spatial Information Sciences. Proceedings ofTheXIX CIPA Symposium. Antalya. XXXIV-5/C15 1682-1777.

42.   Ioannides M, Fellner D, GeorgopoulosA, Hadjimitsis D(2010) 3rd International Conference Euromed2010 Dedicated On Digital Heritage. Limassol, Cyprus.

43.   Cano M, Garzón E, Sánchez-Soto PJ (2013) Historic Preservation, GIS, and Rural Development: The Case of AlmeríaProvince, Spain. ApplGeogr42: 34-47.

44. Tapinaki S, Georgopoulos A, Sellis T (2005) Design Of A Database System For Geometric Documentation. Proceedings oftheCIPA 2005 XX International Symposium, Torino, Italy.

45. Pieraccini M, Guidi G, Atzeni C (2001) 3D Digitizing of Cultural Heritage. J CultHerit2 : 63-70.

46. Guidi G, Beraldin J, Atzeni C (2004) High-Accuracy 3-D Modelingof Cultural Heritage: The Digitizing Of Donatello's "Maddalena". IEEE TImage Process 13 : 370-380.

47. Arias P, Herráez J, Lorenzo H, Ordóñez C (2005) Control of Structural Problems In Cultural Heritage Monuments Using Close-Range Photogrammetry And Computer Methods. ComputStruct83: 1754-1766.

48. MinisterioDe Fomento,GobiernoDe EspañaCentro NacionalDe InformaciónGeográfica(2011).

49. Saaty TL (1988) Mathematical Methods of Operations Research. New York: Dover.

50. Tshipala NN, Coetzee WJL (2012)A Sustainable Adventure Tourism Development Framework ForThathevondo.Journal of Tourismand Hospitality 1:101

CHAPTER 6

Cultural Values and Sustainable Tourism Governance in Bhutan

Kent Schroeder

International Development Institute, Humber College, 3199 Lake Shore Blvd. West, Toronto, ON M8V 1K8, Canada

ABSTRACT

Governance is recognized as a means to promote sustainable outcomes by democratizing the policy process and potentially harmonizing competing policy interests. This is particularly critical for sustainable tourism policy with its multiple sectors and multiple stakeholders at multiple scales. Yet little is known about the kinds of governance processes and instruments that are able to effectively harmonize competing power interests to better balance economic, ecological, and social concerns. This study analyzes the case of Bhutan and its Gross National Happiness (GNH) strategy as it is applied to sustainable tourism policy. Based on semi-structured interviews and focus groups with 57 state and non-state governance actors, it explores whether Bhutan's unique GNH governance framework successfully harmonizes competing interests in the pursuit of sustainable tourism policy. It argues that the implementation of Bhutanese tourism policy is characterized by diverse and unexpected applications of power by multiple policy stakeholders. These complex power dynamics are not shaped in a meaningful way by the GNH governance instruments. Nor are they rooted in a common understanding of GNH itself. While this situation should subvert sustainable tourism policy, a commitment among state and non-state governance actors to a common set of Buddhist-infused cultural values shapes and constrains policy actions in a manner that promotes sustainable tourism outcomes.

Keywords: Bhutan; Buddhism; sustainable tourism policy; governance; Gross National Happiness

1. INTRODUCTION

Sustainable tourism has emerged as a key part of national sustainability policies and strategies. As a concept, however, it continues to face challenges. Despite being an area of focus since the 1980s, research on sustainable tourism has continued to be characterized as a "muddy pool" [1] or as "patchy, disjointed and at times flawed" [2]. One of the key challenges is the issue of balance that is inherent in the concept of sustainable tourism itself. Building on the Brundtland Commission's definition of sustainable development, the United Nations World Tourism Organization (UNWTO) conceptualizes sustainable tourism around this notion of balance:

Sustainable tourism development guidelines and management practices are applicable to all forms of tourism in all types of destinations, including mass tourism and the various niche tourism segments. Sustainability principles refer to the environmental, economic, and socio-cultural aspects of tourism development, and a suitable balance must be established between these three dimensions to guarantee its long-term sustainability [3] (p. 11).

Others confirm the centrality of balancing economic, ecological and socio-cultural systems as a means towards tourism that is sustainable [4,5,6]. What this balance should look like, however, and how it might be achieved through policy remains elusive. Some have argued that the focus of sustainable tourism has privileged environmental sustainability at the expense of other factors [7]. Others argue that sustainable tourism tends to default to a preoccupation with economic growth often couched within neoliberal terms [8,9]. Socio-cultural components, on the other hand, do not receive sufficient attention [10].

The issue of how to better balance economic, ecological, and social considerations in sustainable tourism policy is, therefore, a central one. What is often missed, however, is the connection between this process of balancing the multiple components of sustainability and the competing power relations within which this search for balance exists. This is particularly critical given the broad range of tourism policy stakeholders at multiple scales with potentially competing economic, ecological, and socio-cultural interests. There is a rich literature on the intersection of politics and tourism, generally [11,12,13,14,15,16,17], but only recently has focused attention been paid to the links between governance, sustainable tourism policy, and power dynamics [4,6,18,19]. Governance, defined as the interactions among public, private and civil society actors in the exercise of power and authority, is viewed as a potentially effective means to shape power dynamics and collective action among diverse policy actors to promote sustainable outcomes. It has the potential to enhance democratic decision-making processes and provide tailored policy instruments to harmonize sustainable tourism policy interests [20] (p. 412).

The record of governance practice, however, is less clear. Bramwell [18] (p. 461) argues that the practice of tourism governance has faced difficulties given its focus on multiple policy fields and multiple tourism actors. Hall [4] goes further, stating that the record of tourism governance is largely one of policy failure. As a concept, governance may be key to fostering collective action to promote sustainable tourism policy, but how it might do so in practice continues to require deeper exploration.

This paper addresses this gap through an analysis of Bhutan's unique Gross National Happiness (GNH) strategy and its accompanying governance framework as they are applied to sustainable tourism policy. Bhutan is often portrayed in the popular media as a "last Shangri-la" that has achieved sustainable tourism policy through its GNH approach. Little is known, however, about the actual performance of the GNH governance framework and its influence on sustainability policy. Is it responsible for harmonizing competing policy interests that drive sustainable tourism outcomes or is something else at play? If it is, are its governance instruments replicable elsewhere? If not, what is the engine behind the achievement of sustainable tourism outcomes? Is it even accurate to characterize Bhutan's tourism outcomes as sustainable? This paper explores these questions. It is structured around an analysis of the competing political dynamics of tourism policy implementation, the role of the GNH governance framework in attempting to harmonize these interests in a manner that promotes sustainable balance, and the actual policy outcomes that emerge from the process. The paper argues that Bhutanese policy has achieved notable success in generating sustainable tourism outcomes but the GNH governance tools, while holding significant promise, are not responsible for them. Rather, a common commitment among tourism governance stakeholders to Buddhist-inspired cultural values appears to shape and constrain potentially competing power interests in policy implementation. State and non-state governance actors may have different interests and attempt to impose their own priorities on tourism policy, but these differences are a matter of operational degree rather than kind. They are shaped and harmonized by a common commitment to the Bhutanese cultural values of integration and interdependence that promote balance across economic, social, and ecological systems. Sustainable tourism outcomes are the result.

2. BHUTAN AS A CASE OF TOURISM GOVERNANCE AND SUSTAINABILITY POLICY

Bhutan offers a particularly intriguing case for an exploration of governance and the implementation of sustainable policy. All policy in Bhutan, including tourism policy, is rooted in the country's Gross National Happiness development strategy. GNH is based on the simple notion that happiness should be the foundation for development. Moreover, happiness is a multidimensional and integrated concept, as opposed to the historical focus on development as economic growth. Initiated in the 1970s by the country's fourth king, GNH was originally conceptualized as a synthesis of four interdependent pillars: equitable socio-economic development, environmental conservation, cultural preservation and promotion, and good governance. More recently the pillars have been expanded into nine domains, but the four pillars continue to be recognized as the broad framework through which GNH policy is operationalized [21] (p. 18). Key to the GNH framework is an understanding of the pillars as holistic and integrated. All policy, including tourism policy, must take into account the interrelationships across social, economic, ecological, cultural, and governance systems. This focus on integration is explicitly rooted in Buddhist values [22] (p. 19). Interdependence, harmony, balance, sustainability, and compassion are the value foundations upon which GNH rests as Bhutan's national strategy. Significantly, these values, while Buddhist, are defined as cultural values, embodying the heart of a constructed national Bhutanese culture within a multi-ethnic and multi-religious society. Respondents in this study referred to these cultural values in a variety of ways, including Buddhist values, Buddhist-Hindu values and, most frequently, Bhutanese values.

Tourism policy was initiated in 1974 in Bhutan at roughly the same time as GNH was inaugurated as the country's national development framework. Since then, the implementation of tourism policy has been accompanied by the gradual evolution of a distinct governance framework. Bhutan was an absolute monarchy from 1907–2008. Under the fourth king, absolute power was increasingly devolved and decentralized until democratization in 2008. In terms of tourism governance, this process witnessed a gradual but deepening engagement with a range of state actors, international donors and, increasingly, non-state actors. While this broadens and democratizes policy decision-making, it also opens up the governance process to potentially competing political dynamics. Since democratization in 2008, Bhutan has introduced a set of GNH policy instruments as part of its governance framework that attempt to ensure competing policy interests are shaped and harmonized by the pillars of Gross National Happiness. They require that the multiple dimensions of GNH be incorporated and integrated into all policy at the design, implementation, and measurement stages. The intended result is the sustainable balancing of socio-economic, environmental, cultural, and governance dimensions within all policies.

The GNH Commission is the key GNH governance structure. It is the apex body responsible for the operationalization of GNH in policy design, planning and implementation. Specific policy instruments include the GNH policy selection tool, a draft GNH project selection tool, GNH committees, and the GNH index. The policy and draft project selection tools are instruments that require governance actors to evaluate and rank draft policies and projects against a set of screening questions based on the domains of GNH. The tools ensure that regardless of the policy, it will explicitly take the balancing of the GNH domains into account. GNH Committees are structures meant to exist within each ministry and agency in the central government as well as within sub-national governments. The committees are meant to act as links to the GNH Commission and to ensure that GNH is mainstreamed into policy implementation at all levels of government. The GNH Index is a composite measurement tool that measures policy outcomes in a multidimensional way. It is based on the nine domains of

GNH and includes 33 variables with over 120 indicators. The multidimensional nature of the GNH Index ensures that policy outcomes are measured in a way that is sensitive to sustainability rather than relying solely on GDP per capita. Such multidimensional measurements can then feed back into the policy design and implementation processes. Taken in total, the GNH structures and tools represent a unique set of policy instruments that put the search for balance inherent in sustainability at the heart of the governance and policy process in Bhutan.

Bhutan's evolving governance approach therefore prioritizes a broadened set of state, non-state and donor actors to better democratize sustainable development decision-making while incorporating the unique set of GNH policy tools to ensure broadened decision-making remains structured around GNH's multidimensional approach to sustainability. All of this is rooted in a foundation of cultural values that emphasize interdependence, balance and harmony. Bhutan's GNH approach represents a sustainability governance model for tourism and other policies that seems to hold significant promise. However, does it successfully shape potentially competing political dynamics among tourism policy actors in a manner that generates sustainable tourism outcomes? The following sections turn to an analysis of this question.

3. STUDY METHODS

Semi-structured interviews were undertaken with key informants representing the multiple state and non-state tourism actors in Bhutan. This included government respondents at the central, dzongkhag (district) and gewog (village block) levels; private sector tourism actors; civil society organizations involved in tourism; and international donors. As Bhutan is made up of 20 dzongkhags and 205 gewogs, purposive sampling was used to identify four dzongkhags and 19 gewogs for participation in the research. Both dzongkhags and gewogs were selected to reflect equal representation across the country's four regions and roughly equal distribution across high, medium, and low levels of poverty defined in terms of household consumption.

Semi-structured interviews were supplemented by a number of focus group interviews. The focus groups explored, expanded, and triangulated themes that came out of individual interviews. A general interview guide with open-ended questions was used to frame both the semi-structured interviews and focus group discussions. The guide promoted in-depth discussion of respondents' understandings of GNH, their practices and interactions with other tourism governance actors in the implementation of tourism policy, the nature and impact of various GNH governance instruments on shaping these practices, and the tourism policy outcomes that result.

In total, 57 tourism stakeholders were interviewed through individual or focus group interviews. All respondents and their geographic locations remain anonymous in the study. No personal information was collected for semi-structured interview respondents nor were their names or geographic location shared with other respondents. Personal information was also not collected for focus group participants nor were the names or geographic location of focus group participants shared with other participants in the study.

The interviews and focus groups were further supplemented by analysis of policy documents, site visits to relevant tourism sites and participant observation in a public stakeholder meeting. Interviews and focus group discussions were recorded and transcribed. The transcripts were imported into NVivo software as were field notes from site visits and relevant policy documents. The data were then coded with NVivo through an iterative process that identified individual themes that were aggregated into broader themes related to the GNH intentions of tourism policy, the nature of policy implementation and the resulting tourism outcomes.

4. GROSS NATIONAL HAPPINESS, GOVERNANCE, AND SUSTAINABLE TOURISM POLICY

4.1. Early Years: Monarchy and the Emergence of Sustainable Tourism Policy

Tourism policy was initially formulated and implemented in the Bhutanese political context of the 1970s. The fourth king began his reign as an absolute ruler. The civil service was powerful but its interests generally paralleled those of the king [23] (p. 242), [24] (p. 224). A representative National Assembly existed alongside the monarch and held fairly significant formal power but in practice it deferred to the monarch's political agenda [23] (pp. 243–244), [24] (p. 156). The monarchical regime focused on tourism policy rooted in the GNH pillars. It prioritized the maximization of foreign currency exchange and employment opportunities while simultaneously minimizing the potential negative impacts of this economic growth on traditional culture and the environment. Multiple policy documents outline this policy intention in remarkably consistent terms over time [25] (p. 18), [26] (pp. 67–71), [27] (pp. 16–17). A key tool for balancing the GNH pillars of maximized economic growth and minimized cultural and ecological disruption was, and remains, a daily tariff system. Tourists must take an all-inclusive package tour for which they pay an expensive daily tariff, initially set at $130/day, in foreign currency. The tariff serves multiple cultural, environmental, economic, and social purposes. First, its high daily cost limits the number of tourists who can visit the country as a means to minimize negative environmental and cultural impacts. Second, the high cost also drives economic development despite limiting tourist numbers. Third, a government royalty built into the tariff serves as a means to fund education and health care in the country. In the early years of Bhutanese tourism, this policy approach was termed high value, low volume. It represented the government's priority of balancing the socio-economic, cultural, and environmental pillars of GNH by deriving high economic value through the high cost tariff while simultaneously keeping the volume of tourists low to protect cultural and environmental concerns.

In addition to the tariff, implementing the high value, low volume strategy involved strict regulations around tourist-related cultural and trekking activities. Much of the country in the early years of tourism policy was also closed to tourists. Moreover, the entire tourism experience—marketing, transportation, guiding, accommodations, food—was delivered by the Department of Tourism. This tightly-controlled implementation of tourism policy ensured that economic growth did not overrun cultural and ecological concerns. Indeed, by the late 1980s, concern about growing tourist numbers led to an increase in the daily tariff in order to slow the growth in numbers, yet maintain their economic impact.

Tourism governance evolved in the 1990s. The tourism industry was privatized in a limited manner in 1991 by allowing a relatively small number of private tour companies to replace the government as the front-line providers of tours. Further pressure from the private sector led to full privatization by 1999 that opened up the tourism sector without any limits on the number of tour companies that could register as private sector companies. This resulted in a mushrooming of tour operators to 600 by 2012 [28] (p. 163). Privatization was accompanied by a set of government regulations and a code of conduct to ensure private sector stakeholders engage in their work in a manner consistent with the GNH pillars [29] (pp. 13–37), [30].

The motivation to privatize was an increasing recognition by the monarchical government of the economic potential of tourism and a need to better spread its benefits and employment opportunities [31] (p. 110). Significantly, entry of the private sector into the tourism industry led to the development of civil society organizations (CSOs) to promote private sector interests. The Association of Bhutanese Tour Operators (ABTO) was the first, created in 2000 and later registered as a CSO in 2007. ABTO's creation in 2000 foreshadowed the emergence of increasingly muscular non-state tourism governance actors alongside a growing policy conflict over the appropriate operational balance of the economic pillar with the cultural and environmental pillars of GNH.

4.2. Evolving Policy and Evolving Governance

Changes in Bhutanese society led to a rethinking of the role of tourism policy at the beginning of the 2000s. A gradual move away from an agricultural economy was combined with rising population growth and a growing youth population less interested in agricultural work. Tourism became increasingly viewed as a strategy to address these socio-economic changes. Bhutan 2020, the country's development vision, called for a dramatic increase in tourism revenues by 2017 as a means to promote relevant employment opportunities through economic growth [32] (pp. 26–27). The high value, low volume approach was replaced by high value, low impact. The revised policy approach represents a subtle shift away from minimizing the volume of tourist arrivals to a new emphasis on increasing them to promote greater economic growth, while continuing to keep the negative impact on culture and the environment low. High value, low impact therefore represented a rebalancing of the GNH pillars in tourism policy. Several key government documents outline what this evolution in balance looks like [25,26]. On the one hand, greater focus on economic growth through increased tourist arrivals still requires protecting Bhutanese culture and the environment from the potential excesses of such growth. On the other hand, protecting traditional culture and the environment is also a means to promote Bhutan as a unique niche tourism destination as fuel for economic growth. Cultural and environmental concerns are therefore to be protected from economic growth while simultaneously driving it. Within this apparent contradiction is a potentially virtuous sustainability circle where GNH pillars mutually reinforce one another: marketing a protected culture and environment will attract increased numbers of tourists which, in turn, requires further protection of the country's culture and environment to continue to promote Bhutan as an exotic tourism destination. Sustainability is still the goal but requires a rebalancing of the GNH pillars to further emphasize growth to address changing economic conditions while still maintaining the integrity of the environmental and cultural pillars.

The delicate balance inherent in the high value, low impact strategy shift was accompanied by a further evolution in tourism governance. Consistent with the good governance pillar of GNH, the new approach promoted broadened and deepened participation by communities, dzongkhag, or district, level governments and the private sector. A number of new tourism related civil society organizations also emerged in addition to ABTO. The Handicrafts Association formed in 2005, the Hotel Association in 2008 and the Guide Association of Bhutan (GAB) in 2009. In this context of broadening actors, the Department of Tourism was replaced in 2008 with the Tourism Council of Bhutan (TCB). The creation of TCB demonstrated an attempt to further broaden participation in sustainable tourism governance, at least in theory. Described by a senior official within TCB as "a governance experiment" and "a bold step", it is a governance model based on the recognition of the multi-sectoral nature of sustainable tourism that requires broader stakeholder participation. TCB is an autonomous council of state and non-state stakeholders. It brings together representatives of multiple central government ministries, autonomous commissions, the private sector, and civil society organizations. It is chaired by the Prime Minister and is mandated to formulate and implement tourism policy, develop regulations, diversify tourism products, and lead tourism human resources development.

The expansion of tourism governance actors through TCB to incorporate non-state actors would appear to democratize the process of implementing the high value, low impact policy shift. It also opens up this democratic space to potentially competing priorities among state and non-state governance actors that may undermine sustainable tourism outcomes, a situation for which the GNH policy tools were developed to address. The following two sections explore how this dynamic has played out in practice in the interactions across sub-national governments and TCB as well as between the private sector, CSOs, and TCB.

4.3. Sub-National Governments and Tourism Policy

The move to the high value, low impact strategy involved, among other aspects, increasing the involvement of dzongkhag-level governments and administrations in the implementation of tourism policy. Of the four dzongkhags analyzed in this study, diverse applications of power and inconsistent roles were evident, particularly among the two dzongkhags with significant tourism activity. In one case, the dzongkhag played a critical role in the development and implementation of new tourism products within the district. The Dzongdag, or chief executive of the dzongkhag administration, stated that "TCB helps us and provides funding, but we are in control of what tourism will look like". TCB's supporting role in this case is providing marketing, funding and training where appropriate. In contrast, the other dzongkhag with significant tourism activity was characterized by almost no involvement by the dzongkhag administration. Officials stated that their priorities are elsewhere, particularly as private sector actors and TCB, in their view, maintain the most power in implementing tourism policy. According to a Dzongkhag Planning Officer, "[tourism] is driven by the tour operators and there is no need for us". As a result, TCB and private tour operators tend to dominate the process of tourism policy implementation in this dzongkhag.

At the gewog, or village block, level, a somewhat different situation exists. The decentralization of implementing tourism policy does not significantly go beyond the dzongkhag level. Yet local government administrations at the gewog level again have different patterns of power applied to local tourism activities. In most instances, gewog officials play a very limited role. Some play no role at all. In the case of planning a tourist festival to be held in his gewog, a gewog official flatly stated "we make the seating arrangements; no planning". In a few cases, however, gewog officials are freelancing outside of official channels in developing and implementing local tourism activities. According to one elected gewog official, there are "too many formalities working with them [TCB] so I am exploring these things on my own". This local official represents several who are moving ahead and implementing unofficial tourism activities with TCB playing no role. The case of freelancing gewogs is significant. While limited in number, these initiatives, unconstrained by official channels, open up the opportunity for local tourism initiatives that do not reflect the GNH balance intended in official tourism policy. They open up the opportunity for unsustainable practices as local governments apply power despite it not being formally granted to them.

Several things are notable about the involvement of sub-national levels of government. First, the engagement of lower levels of government, where it occurs, is not structured in any meaningful way by the relevant GNH policy tools like the GNH Committees, draft GNH project selection tool and the GNH Index. Again, these tools are meant to ensure that the multidimensional and sustainable focus of GNH is incorporated into all aspects of the tourism policy process as a means to harmonize potentially competing power interests. Sub-national state actors, however, tend to be unaware of their existence or unconvinced of their usefulness. A freelancing gewog official, for example, stated "A GNH committee does not exist in the gewog. I tell the [village representatives] we just need to work as one, with equality we serve the benefit of the people". Similarly, a high ranking dzongkhag official claimed, "I don't think we need any GNH committee as it is a philosophy which everybody is aware of and where everybody is involved in this. So I don't know what work that committee would do".

The second notable issue related to the involvement of sub-national governments is that many officials at the dzongkhag and gewog levels maintain different understandings, and often misunderstandings, of the very nature of GNH itself. Many officials are quite open that they simply do not understand GNH despite the fact that they are charged with implementing it as part of Bhutan's national development strategy, including tourism. One official at the gewog level, embarrassed by his lack of understanding of GNH, sheepishly admitted, "I know I should know what GNH is, but I don't. I think it has something to do with four pillars". A high ranking dzongkhag official was more blunt: "Even I am confused about GNH". There are multiple causes for the confusion around GNH among these governance actors. Often it relates to the GNH Index measurement tool with its expansion from four GNH pillars to nine GNH domains with 33 variables and over 120 indicators. One of the GNH policy tools itself has therefore muddied a clearer understanding of the larger strategy.

The diverse applications of power across sub-national governments combined with misunderstandings of GNH and the lack of use of its policy tools would seem to undermine the implementation of tourism policy that sustainably balances the GNH pillars. It opens the possibility for different sub-national state actors to impose inconsistent and potentially competing interests on the implementation of tourism policy unconstrained by GNH. However, this is not the case. Multiple dzongkhag officials indicate that they do not use the GNH tools because their decisions and actions are naturally structured by a common set of Buddhist-based cultural values—interdependence, harmony, balance, and sustainability. According to one official, "… the system [GNH tools] is not there…. It [GNH] is just done because Bhutanese have a set of values focused on that". For these dzongkhag officials, they are already doing what the GNH tools intend to do as it is an inherent part of their value system. Similarly, dzongkhag officials that demonstrate a misunderstanding of the nature of GNH also speak of how they implement tourism policy based on the same cultural values that are the foundation of GNH. Without understanding GNH, their decisions are structured in a way that is consistent with it. This extends to those gewog officials who are freelancing by implementing their own tourism activities outside of official channels. In these cases, their unofficial tourism activities clearly target a balance of economic, cultural, and environmental concerns. In describing his unofficial tourism plans, one elected local official stated, "our culture is also very traditional. We want to attract tourists in a way that does not affect tradition and architecture and environment in the valley". These officials again, unprompted, describe their cultural values, rooted in Buddhism, as the driver of this. While they do not do this in reference to the GNH pillars, the values they describe and the influence on their actions is again consistent with GNH. An intriguing situation therefore exists. The GNH governance framework has had some success in broadening the input of sub-national government actors in implementing tourism policy yet has failed to engage these same officials in using the GNH policy tools as a means to promote sustainable tourism outcomes. At the same time, a common commitment among these officials to a common set of cultural values—the same values that underlie GNH—structures their actions in implementing sustainable tourism policy.

4.4. Non-State Actors and Tourism Policy

The increasing engagement of sub-national governments in the high value, low impact strategy is paralleled by an intention by government to more broadly engage non-state actors in tourism governance. Complicating this situation was the 2009 recruitment by the central government of McKinsey and Company, an international consulting company, to assist with the accelerated implementation of a range of policies, including tourism policy. McKinsey moved forward in partnership with TCB and other stakeholders to implement the greater focus on economic growth. One of the strategies suggested by McKinsey was controversial among many private sector and CSO governance actors. The daily tariff and required tour package have always been the hallmarks of tourism policy in Bhutan. However, in order to put greater focus on the economic development pillar of GNH, McKinsey suggested a complete liberalization of the tourism industry with a target of 250,000 tourist arrivals by 2013, approximately a ten-fold increase from tourist numbers in 2009 when McKinsey made the proposal. Liberalization would occur by dropping the tariff and package tours in an effort to dramatically increase the number of tourists to fuel accelerated economic growth. Government respondents were clear that this was rooted in the need to rebalance economic, environmental and socio-cultural concerns as the Bhutanese economy evolves in the context of increased urbanization and youth unemployment. Changing conditions required changing the balance across the GNH pillars. The values underlying GNH, according to these respondents, drive this rebalancing.

The proposal was approved by the prime minister and cabinet. Some TCB members from the private sector and civil society complained that they were not consulted in any meaningful way despite being a part of TCB. One stated that "they often turn a deaf ear towards us". Another suggested "they do listen, but whether they take us seriously is debatable". A third was more blunt, stating TCB was "trying to play a dirty game". Moreover, many private sector and civil society representatives disagreed with the proposal. Their reasons for opposition are intriguing. Some tour operators opposed it as they perceived it as a strategy that would cut into their profits now guaranteed through the daily tariff. Others, however, made a much more subtle argument rooted in competing perceptions of how the GNH pillars should be balanced in practice. The majority of private sector actors interviewed opposed McKinsey's proposal because it was perceived as accelerating the economic pillar of GNH at the expense of sustainability. "TCB is too focused on numbers" and "TCB always has an agenda that is just based on profit" were common comments. One respondent went further, stating that the policy pivot to increased focus on economic growth "frightens me". Many outlined their concern that Bhutan's cultural heritage would be eroded or its pristine environment would suffer. One tour operator, referencing the impacts of a liberalized tourism policy in a neighbouring country, stated, "as Bhutanese, we feel so much for preservation of culture and identity. If we liberalize we'll be no different than Nepal". At first glance, this argument sounds counter-intuitive coming from the private sector. Yet many respondents demonstrated a nuanced understanding of the interconnectedness among economic, cultural and ecological issues. By dropping the tariff as a means to dramatically increase tourist arrivals to 250,000, many argued that the subsequent erosion of traditional culture and the environment, unwanted on their own, would in turn also decrease the attractiveness of Bhutan as an exotic tourist destination, threatening the future economic potential of Bhutan's tourism industry. According to one tour company operator: "Profit is not everything. Our philosophy and belief is: if we are profitable as a society, as a community, as a tour company, we need to take care of these [cultural and ecological] things. If not we'll kill the golden goose". Another claimed that environmental and cultural damage of 250,000 tourists would "destroy the whole image of Bhutan" as a niche tourist destination. A third suggested that the perceived overemphasis on economic growth in McKinsey's proposal will lead to "the final outcome where tourism will be a messed up business in the country". Respondents from non-state organizations made this case, often vigorously and emotionally, with an appeal to Buddhist-inspired cultural values of interdependence and harmony. The need for better balance across economic, environmental, and cultural systems was argued as being a foundational part of the Bhutanese value system. McKinsey's proposal was viewed as a threat to that balance.

Again, some non-state actors made this case with explicit reference to GNH while others made no reference to GNH or claimed not to understand it. In the case of the former, some claimed the government's position was abandoning GNH, leaving the private sector to take up the cause. "Even though GNH is supported by government," stated one private sector respondent, "who is doing it? So we have to try". In the latter case, several respondents made implicit GNH arguments while claiming not to understand it. For example, a representative of a civil society organization who opposed the increased economic focus of TCB's strategy argued for better integration of economic, cultural, and environmental considerations based on Buddhist values. At the same time, he also claimed "GNH is too complicated for us... It is a good philosophical guide but I don't really understand it". This respondent was making a GNH argument without realizing it.

The opposition of many non-state governance actors to McKinsey's proposal demonstrates a debate over the proper balancing of GHN pillars in practice in order to pursue a sustainable tourism path. The position of McKinsey and TCB prioritized greater economic growth in order to re-balance the GNH pillars to promote sustainability in light of changing economic and social conditions in Bhutan. Those in opposition, however, viewed the strategy as not a re-balancing of the GNH pillars but an unbalancing that undermines sustainability. The same set of cultural values structured both positions.

The different perspectives on how to move forward with a sustainable tourism policy took a dramatic turn in 2010. A meeting of tourism stakeholders involved the prime minister, as the chair of TCB, outlining the new direction under McKinsey's proposal. While some non-state stakeholders supported the strategy or remained neutral, many opposed it vocally. Most intriguingly, behind the scenes, respondents from the Department of Forests and Parks Services (DoFPS) supported these non-state actors in their opposition to the strategy. This represented a delicate balancing act as DoFPS officials disagreed with the strategy yet, as government officials themselves, felt it inappropriate to publicly oppose the government. The alternative was to offer moral support to the private sector actors opposed to McKinsey's proposal. According to one DoFPS official: "Personally, from our side, because I know many tour operators, we exchanged a lot of dialog and said that 'you guys go against the government.' For us, we work in government and it is very difficult for us". The DoFPS respondents emphasized that they supported the private sector given their concerns over the policy shift's implications for the upending of the balance across economic, environmental, and cultural concerns. The expanded set of tourism governance actors therefore demonstrated a diverse and fractured set of power relations rooted in differences over the appropriate balance of GNH pillars. An official government position rooted in GNH was supported by some non-state actors while an opposing set of non-state actors promoted an alternative GNH argument supported by another government entity.

Faced with a group of stakeholders opposing the policy direction using an alternative argument on sustainability, the Prime Minister reversed the proposal despite cabinet approval. Not only was the daily tariff maintained, it was increased to $250/day for high tourist season and the target number of new tourist arrivals was decreased from 250,000 to 100,000. This reversal represented a significant triumph to many respondents within the private sector and civil society. One private tour operator stated "we had a better case than

McKinsey who were paid US$9 million by the government". A policy change rooted in GNH—accelerating economic growth through increased tourist numbers to address changing socio-economic conditions—was overturned based on opposition that also appealed to the GNH pillars or values in a more integrated way. The policy conflict was not over broader tourism policy or GNH itself, but how the economic, cultural and environment pillars of GNH should be balanced in practice to best promote sustainability. It represented a dynamic process of finding the appropriate balance of GNH pillars where an expanded set of governance actors differed over the operational nature of the balance but whose differences were constrained by common cultural values that prioritize sustainable balance.

Intriguingly, the issue of removing the tariff arose again in late 2015. The National Council, the upper house in Bhutan's parliament, recommended after a year-long review to maintain the $65 government royalty component of the $250 tariff but remove the rest of it. Doing so, it argued, would address the emerging problem of tour operators undercutting the cost of the tariff by charging tourists less and would also address the perceived unequal distribution of tourism benefits across the country [33]. The argument was justified as a means of ensuring the core GNH values underlying the high value, low impact strategy are not undermined [33]. While no final decision has yet been made at the time of writing, the recommendation was immediately met with opposition. The opposition was again rooted in an argument of what the proper balance of economic, social, cultural, and environmental concerns should look like [34]. The same pattern therefore emerges: differences arise over the operationalization of balancing GNH pillars in tourism policy but the issue of interdependent balance itself is not disputed.

4.5. Cultural Values: Constraining and Shaping Policy Actions

The analysis above suggests that the implementation of sustainable tourism policy in Bhutan is characterized by diverse and fractured expressions of power. Governance actors apply or try to apply their interests to the policy implementation process in different ways in different contexts. Some sub-national governments successfully influence tourism policy while others do not; non-state actors engage with policy implementation through both cooperation and opposition to the government. Overall, the democratization of Bhutan's tourism governance framework has led to a situation where both state and non-state actors impact the direction of policy implementation. Within this context, two issues remain significant. First, the GNH specific policy tools are notable for their on-going absence in shaping how power is applied by governance actors in the implementation of tourism policy. These tools are meant to ensure that the multidimensional and sustainable focus of GNH is incorporated into all aspects of the tourism policy process. While debate among governance actors occurred on the proper balance of GNH pillars, this was not structured by the GNH policy tools as should be the case. Indeed, some government officials were highly ambivalent about the tools. Second, and even more curious, is that while many governance actors framed their positions in terms of how to best balance the GNH pillars, others did not. Multiple governance actors, both state and non-state, did not understand the nature of GNH itself. In these cases, GNH arguments were often being made without understanding their connection to GNH. A common commitment to cultural values appears to fill this void. Whether national government, sub-national government or non-state actors, a commitment to these values—balance, harmony, moderation, interdependence—structures governance actors' perspectives and actions whether or not they understand GNH and its governance tools. Differences over the balance of the economic, cultural, and environmental GNH pillars are therefore a matter of operational degree rather than kind. Fractured expressions of power are shaped and constrained by cultural values in the policy implementation process. The remaining question, though, is whether or not this is enough to generate sustainable tourism policy outcomes.

5. TOURISM POLICY OUTCOMES

An analysis of the outcomes of tourism policy suggests that the fractured process of policy implementation constrained by cultural values has had notable success in achieving sustainable outcomes consistent with the initial Gross National Happiness policy intentions. This is despite the lack of use of GNH tools or common understanding of GNH itself. Economic growth has been balanced by successes in environmental and cultural protection. Bhutan received 274 international tourists in 1974 when tourism policy began [26] (p. 15). This number rose significantly after privatization of the industry in the 1990s with 7158 tourist arrivals in 1999 and 44,252 in 2013 [35,36]. This increase is dramatic within the Bhutanese context but it pales in comparison to some of its South Asian neighbours. Maldives, for example, a country with just over half the population of Bhutan and a fraction of Bhutan's area, received over one million tourists in 2013 [37]. The relatively small numbers of tourists in Bhutan, however, have generated significant economic outcomes given the high cost of the daily tariff. Tourism is the largest contributor of foreign currency in Bhutan [36] (p. 174). It has regularly fueled double digit percentage growth in government revenues [28]. Tourism related employment has also increased. The number of direct and indirect jobs in the industry reached 21,289 in 2010, representing significantly more than the government's goal of 18,000 jobs [38] (p. 41). By 2012, this increased to 28,982 jobs [39]. Significant challenges still remain with many seasonal jobs and a small percentage of tour operators dominating the industry [28] (p. 163). The sustainability of the tourism industry also remains susceptible to international influences beyond Bhutan's control. For example, the ongoing political unrest in Thailand and the 2015 earthquake in Nepal, two key access points for travelling to Bhutan, have dampened tourist arrivals [40]. Overall, however, tourism policy has been fairly successful in generating the intended economic outcomes.

Economic outcomes have been balanced by successes in environmental protection. Again, the GNH driven intention of tourism policy is to use a high tariff to promote economic growth while continuing to limit the negative impacts on Bhutan's environment and culture. Trekking represents the biggest threat to Bhutan's environment. Nepal's experience is particularly notable given the deforestation, trail erosion, contamination of water, and solid waste problems arising from tourist trekkers [41,42,43,44]. Bhutan's relatively small number of high paying tourists combined with strong regulations around garbage and fuel wood have helped avoid this problem. There is some evidence of the destruction of vegetation for firewood and minor erosion of trekking trails but deforestation has not been a significant issue [31] (p. 120), [45] (p. 499). Garbage on trekking trails has historically not been a problem [31] (p. 120) although it is an increasing concern, particularly as more packaged foods are available from India.

The economic and environmental outcomes of tourism policy have been balanced by the protection and promotion of Bhutanese culture. Given that most tourists visit Bhutan for a cultural experience, traditional culture is perhaps most at risk of erosion from tourism. Again, however, cultural outcomes associated with tourism, like environmental outcomes, have a relatively good record. The intention of tourism policy is to both protect Bhutan's culture from the excesses of tourism while actively promoting it as an exotic experience to attract tourists. In this context, the promotion of culture for tourism purposes has successfully uncovered latent cultural practices that had disappeared from Bhutanese life [31] (p. 259), [46,47]. Moreover, relatively little erosion of traditional culture has occurred. Many of the problems that have emerged in tourist destinations with mass tourism—begging, theft, sex tourism—are not evident in Bhutan [31] (pp. 121–122), [46] (p. 259). This is not to suggest that there are no challenges to Bhutanese tradition culture. Evidence exists of religious and cultural practices being monetized or changed to attract tourist dollars as an emerging sense of materialism develops [47]. Given the overall intent of the policy, however, Bhutan has achieved significant success in protecting traditional cultural practices and uncovering latent ones.

The balanced economic, ecological, and socio-cultural outcomes of tourism policy suggest that the cultural values that structure the perceptions and actions of tourism policy actors play an important governance role. At the same time, a significant issue intrudes. Regional tourists from India, Bangladesh, and Maldives are exempted from the policy. They do not require the daily tariff and tour package. The vast majority of these regional tourists come from India. For years, accurate figures for regional tourists did not exist. More recently, better tracking found a yearly figure of 63,426 regional tourists, almost half the number of overall international tourists [36] (p. 176). What can be made of this? It would seem to be inconsistent with the GNH intentions of Bhutan's tourism policy. Being exempt from the daily tariff and package tours suggests increasing numbers of regional tourists may flood the Bhutanese tourism market undermining the ecological and cultural traditions Bhutan's tourism policy has historically focused on protecting. The issue is a complicated one, particularly in the case of India which is a dominant trade and geopolitical partner for Bhutan.

The key to this apparent contradiction can likely be found in the nature of Gross National Happiness. GNH is a sustainability strategy but it is also a geopolitical strategy. It was constructed as a national strategy that integrates the four pillars as a multidimensional sustainability framework and locates this framework as a key part of the identity of the Bhutanese state itself. The uniqueness of this national identity is meant to differentiate and protect Bhutan as a sovereign entity in a region where sovereignty has not always been maintained. Past regional experience where sovereignty was lost, subsumed or threatened in Tibet, Sikkim, and Darjeeling has had a powerful influence on Bhutan's leaders [48] (p. 36). At the same time, the uniqueness that GNH bestows on Bhutan as a sovereign nation still requires that the country navigate the geopolitics of the region. Accordingly, the cultural values of balance and compromise that underlie GNH do not separate economic, ecological, or cultural concerns from issues of state sovereignty in a region of geopolitical giants. In fact, it is quite the opposite. The maintenance of the Bhutanese state itself is part of the larger sustainability equation [49,50]. Compromise in geopolitical issues with its neighbours is a part of this. Policy compromise may therefore be a necessary component of GNH balance in the context of geopolitical concerns.

None of this changes the potential threat to ecological and cultural concerns that increasingly unregulated regional tourists represent. Their ability to travel without restrictions may contribute to cultural and environmental disruption. However, it also does not suggest that future problems with sustainability are inevitable as regional tourists, and Indian tourists in particular, travel to the country. Indeed, Bhutan's experience with McKinsey illustrates that actions by foreign actors that appear to stray from the required balance of GNH are checked by those who feel the cultural values of integration, balance, and harmony are being violated. This already appears to be emerging with the increase in Indian tourists. Multiple state, private sector, and civil society respondents indicated the need to address the issue now to maintain sustainability. The common set of values that has enabled a dynamic and on-going balancing of GNH pillars in the implementation of tourism policy, even when this involved clashes over operational issues, suggests that new external challenges do not automatically represent a threat to the sustainability balance of GNH.

6. CONCLUSIONS

Governance holds the potential to broaden and democratize decision-making to promote sustainable policies. Challenges still remain, however, in understanding how it can best navigate potentially competing power dynamics among an expanded set of policy stakeholders. Bhutan's experience with its GNH approach to tourism policy and governance offers several insights. It demonstrates that while opening up the governance process to multiple state and non-state actors broadens and democratizes the decision-making structure, it also can generate competing interests that lead to diverse and fractured expressions of power as different actors attempt to imprint their priorities on the policy process in different ways. Sub-national governments in Bhutan either took the lead in implementing tourism policy, abrogated responsibility to TCB, or freelanced outside of official channels. Non-state actors engaged with the state as part of TCB but also opposed TCB, supported by another government agency, over how the GNH pillars should be properly balanced. This is not unexpected. Governance is a complex process of interacting stakeholders that drives often unexpected and emergent priorities and alliances [51]. What Bhutan uniquely offers in this context is its GNH policy tools. Unfortunately little can be said about these instruments and their applicability elsewhere given their general lack of use by tourism governance actors. This is a missed opportunity. The existence of diverse expressions of power demonstrates that policy implementation interactions are complex. The GNH tools were constructed to be sensitive to this complexity and the need for integration. The Bhutanese government needs to institute the use of these instruments on a much broader scale in the process of policy implementation. Without a concerted effort to better apply them in practice, their ability to shape fractured expressions of power to promote sustainable policy is largely unknown. The need to do so has further immediacy given the lack of a clear understanding of GNH itself among many governance actors.

This situation—fractured expressions of power, unused GNH policy tools, contested understandings of GNH itself—should have dire consequences for achieving sustainable tourism outcomes rooted in GNH. Tourism outcomes illustrate that this is actually not the case. Outcomes consistent with GNH are generally being achieved. A key finding emerging from the Bhutanese case is the central role of Buddhist-inspired cultural values in this situation. The values of interdependence, harmony and balance continue to shape policy actors' decisions and actions, even when they disagree over the operational nature of the sustainable balance of GNH pillars. The GNH policy tools may be unused and GNH not commonly understood, but a common commitment to a common set of values helps fill this void and promote sustainable balance. There is a degree of dynamism as governance actors often disagree over the specific nature of the balance in practice, but the necessity of the balance itself is not questioned given its roots in a common set of cultural values. This looks quite different than the policy context in the West where the economy and environment are often pitted as opposing interests.

The role of cultural values in governance rests uncomfortably within the larger development and governance literature. Cultural explanations are often marginalized or, at best, treated as a poor cousin that merely shapes structural, institutional or public choice explanations [52,53,54]. The Bhutanese case suggests something different. It suggests that cultural values not only matter to governance and sustainability, they can matter significantly in promoting sustainable policies by shaping and harmonizing policy actions in the context of fragmented applications of power. This is not to suggest that cultural values always matters. Nor does it suggest that cultural values do not evolve and change. However, it does suggest that cultural values be taken seriously on their own terms as a key component that drives the success of sustainability policy. The challenge, of course, is identifying which cultural values are able to shape governance to promote sustainable policy. Is it restricted to Buddhist-inspired cultural values rooted in interdependence and balance? Are Western values necessarily a barrier to sustainability? The Bhutanese case does not answer these questions but it points to the need for future research to treat cultural values as a potentially key vehicle in the promotion of sustainable policy outcomes.

ACKNOWLEDGMENTS

This work was carried out with the aid of a grant from the International Development Research Centre, Ottawa, Canada. Information on the Centre is available on the web at idrc.ca.

REFERENCES

1. Harrison, D. Sustainability and tourism: Reflections from a muddy pool. In Sustainable Tourism in Islands and Small States; Briguglio, L., Archer, B., Jafari, J., Wall, G., Eds.; Pinter: London, UK, 1996; pp. 69–89.

2. Liu, Z. Sustainable tourism development: A critique. J. Sustain. Tourism 2003, 11, 459–475.

3. United Nations Environment Programme and United Nations World Tourism Organization. Making Tourism More Sustainable: A Guide For Policy Makers; UNEP: Paris, France; UNWTO: Madrid, Spain, 2005.

4. Hall, C.M. Policy learning and policy failure in sustainable tourism governance: From first- and second-order to third-order change? J. Sustain. Tourism 2011, 19, 649–671.

5. Sharpely, R. Tourism and sustainable development: Exploring the theoretical divide. J. Sustain. Tourism 2000, 8, 1–19.

6. Teo, P. Striking a balance for sustainable tourism: Implications of the discourse on globalization. J. Sustain. Tourism 2002, 10, 459–474.

7. Neto, F. A new approach to sustainable tourism development: Moving beyond environmental protection. Nat. Res. Forum 2003, 27, 212–222.

8. Buckley, R. Sustainable tourism: Research and reality. Ann. Tourism Res. 2012, 2, 528–546.

9. Puhakka, R.; Saarinen, J. New role of tourism in national park planning in Finland. J. Environ. Dev. 2013, 22, 411–434.

10. Overton, J. Sustainable development and Pacific islands. In Strategies for Sustainable Development: Experience from the Pacific; Overton, J., Scheyvens, R., Eds.; Zed Books: London, UK, 1999; pp. 1–15.

11. Britton, S.G. The political economy of tourism in the third world. Ann. Tourism Res. 1982, 9, 331–358.

12. Hall, C.M. Tourism and Politics: Policy, Power and Place; John Wiley & Sons: Chichester, NY, USA, 1994.

13. Mosedale, J.T. Political Economy of Tourism: A Critical Perspective; Routledge: New York, NY, USA, 2011.

14. Oppermann, M. Tourism space in developing countries. Ann. Tourism Res. 1993, 20, 535–556.

15. Nunkoo, R.; Smith, S. Political economy of tourism: Trust in government actors, political support, and their determinants. Tourism Manag. 2013, 36, 120–132.

16. Jóhannesson, G.T., Ren, C., van der Duim, R., Eds.; Tourism Encounters and Controversies: Ontological Politics of Tourism Development; Ashgate: Surrey, UK, 2015.

17. Zhang, C.; Decosta, P.; McKercher, B. Politics and tourism promotion: Hong Kong's myth making. Ann. Tourism Res. 2015, 54, 156–171.

18. Bramwell, B. Governance, the state and sustainable tourism: A political economy approach. J. Sustain. Tourism 2011, 4–5, 459–477.

19. Scheyvans, R. The challenge of sustainable tourism development in the Maldives: Understanding the social and political dimensions of sustainability. Asia Pacific Viewpoint 2011, 52, 148–164.

20. Bramwell, B.; Lane, B. Critical research on the governance of tourism and sustainability. J. Sustain. Tourism 2011, 4–5, 411–421.

21. GNH Commission. Tenth Five Year Plan 2008–2013. Volume 1: Main Document; GNH Commission: Thimphu, Bhutan, 2009.

22. Planning Commission. Bhutan 2020: A Vision for Peace, Prosperity and Happiness. Part I; Planning Commission, Royal Government of Bhutan: Thimphu, Bhutan, 1999.

23. Mathou, T. The politics of Bhutan: Change in continuity. J. Bhutan Stud. 2000, 2, 250–262.

24. Rose, L. The Politics of Bhutan; Cornell University Press: Ithaca, NY, USA, 1977.

25. Department of Tourism. Bhutan National Ecotourism Strategy; Department of Tourism, Royal Government of Bhutan: Thimphu, Bhutan, 2001.

26. Department of Tourism. Sustainable Tourism Development Strategy; Department of Tourism, Royal Government of Bhutan: Thimphu, Bhutan, 2005.

27. Royal Government of Bhutan. Economic Development Policy of the Kingdom of Bhutan, 2010; Royal Government of Bhutan: Thimphu, Bhutan, 2010.

28. Royal Monetary Authority. Annual Report 2010/2011; Royal Monetary Authority of Bhutan: Thimphu, Bhutan, 2012.

29. Association of Bhutanese Tour Operators. Tourism Reference Kit. Available online: http://www.abto.org.bt/publications/ (accessed on 24 May 2014).

30. National Environment Commission. Environmental Code of Practice: Tourism Activities; National Environment Commission, Royal Government of Bhutan: Thimphu, Bhutan, 2004.

31. Rinzin, C.; Vermeulen, W.; Glasbergen, P. Public perceptions of Bhutan's approach to sustainable development in practice. Sustain. Dev. 2007, 15, 52–68.

32. Planning Commission. Bhutan 2020: A Vision for Peace, Prosperity and Happiness. Part II; Planning Commission, Royal Government of Bhutan: Thimphu, Bhutan, 1999.

33. Palden, T. Council recommends revision in tourism tariff, Kuensel. Available online: http://www.kuenselonline.com/council-recommends-revision-in-tourism-tariff/ (accessed on 25 November 2015).

34. Norbu, P. Bhutan better off with the current daily tourist tariff Part II. Available online: http://www.kuenselonline.com/bhutan-better-off-with-the-current-daily-tourist-tariff-part-ii/ (accessed on 30 November 2015).

35. Royal Monetary Authority. Annual Report 2004/2005; Royal Monetary Authority of Bhutan: Thimphu, Bhutan, 2006.

36. Royal Monetary Authority. Annual Report 2013/2014; Royal Monetary Authority of Bhutan: Thimphu, Bhutan, 2014.

37. Ministry of Tourism, Republic of Maldives. Tourism Yearbook 2014; Ministry of Tourism, Government of the Republic of Maldives: Malé, Republic of Maldives, 2014.

38. Royal Government of Bhutan. The Third Annual Report of Lyonchhen Jigmi Yoeser Thinley to the Seventh Session of the First Parliament on the State of the Nation; Royal Government of Bhutan: Thimphu, Bhutan, 2011.

39. Tourism Council of Bhutan. Bhutan Tourism Monitor Annual Report 2012; Tourism Council of Bhutan: Thimphu, Bhutan, 2012.

40. Kina, D. Peak Season Records Fall in Tourist Arrivals, Kuensel. Available online: http://www.kuenselonline.com/peak-season-records-fall-in-tourist-arrivals/ (accessed on 26 September 2015).

41. Garbarino, M.; Lingua, E.; Marzano, R.; Urbinati, C. Human interactions with forest landscape in the Khumbu valley, Nepal. Anthropocene 2014, 6, 39–47.

42. Byers, A. Contemporary human impacts on subalpine and alpine ecosystems of the Hinku Valley, Makalu-Barun National Park and Buffer Zone, Nepal. Himalaya 2014, 33, 25–41.

43. Manfredi, C.; Flury, B.; Vivlano, G.; Thakuir, S.; Khanal, S.; Jha, R.; Maskey, R.; Kayastha, R.; Kafle, K.; Bhochhlbhoya, S.; et al. Solid waste and water quality management models for Sagarmatha National Park and Buffer Zone. Mountain Res. Dev. 2010, 30, 127–142.

44. Nyaupane, G.P.; Lew, A.; Tatsugawa, K. Perceptions of trekking tourism and social and environmental change in Nepal's Himalayas. Tourism Geogr. 2014, 16, 415–437.

45. Gurung, D.B.; Seeland, K. Ecotourism in Bhutan: Extending its benefits to rural communities. Ann. Tourism Res. 2008, 35, 489–508.

46. Brunet, S.; Bauer, J.; DeLay, T.; Tshering, K. Tourism development in Bhutan: Tensions between tradition and modernity. J. Sustain. Tourism 2001, 9, 243–263.

47. Reinfeld, M.A. Tourism and the politics of cultural preservation: A case study of Bhutan. J. Public Int. Affairs 2003, 14, 1–25.

48. Banki, S. Resettlement of the Bhutanese from Nepal: The durable solution discourse. In Protracted Displacement in Asia; Adelman, G., Ed.; Ashgate: Aldershot, UK, 2008; pp. 29–58.

49. Wangchuck, J.S. To Special Commission. In Portrait of a Leader: Through the Looking-Glass of His Majesty's Decrees; Nishimizu, M., Ed.; Centre for Bhutan Studies: Thimphu, Bhutan, 2008; p. 7.

50. His Majesty the King's Madhavrao Scindia Memorial Lecture. Available online: http://no.dou.bt/2009/12/23/king-scindia-lecture/ (accessed 19 September 2015).

51. Duit, A.; Galaz, V. Governance and complexity—emerging issues for governance theory. Governance 2008, 21, 311–335.

52. Acemoglu, D.; Robinson, J. Why Nations Fail; Crown Business: Ney York, NY, USA, 2012.

53. Diamond, J. Romney hasn't done his homework. New York Times, 1 August 2012. Available online: http://www.nytimes.com/2012/08/02/opinion/mitt-romneys-search-for-simple-answers.html (accessed on 19 September 2015).

54. Sachs, J. The End of Poverty: Economic Possibilities for Our Time; Penguin: New York, NY, USA, 2005.

CHAPTER 7

Local Residents' Attitude toward Sustainable Rural Tourism Development

Iulia C. Muresan [1,*], Camelia F. Oroian [1], Rezhen Harun [2,*], Felix H. Arion [1,*], Andra Porutiu [1], Gabriela O. Chiciudean [1], Alexandru Todea [1] and Ramona Lile [3]

[1] Department of Economic Sciences, University of Agricultural Sciences and Veterinary Medicine Cluj-Napoca, 3-5 Manastur Street, Cluj-Napoca 400372, Romania

[2] Department of Agribusiness and Rural Development, Faculty of Agricultural Sciences, University of Sulaimani, Kurdistan Regional Government-Iraq, Sulaimani-Bakrajo 5100, Iraq

[3] Department of Economics, Faculty of Economic Sciences, "Aurel Vlaicu" University of Arad, 77 Bdul. Revolutiei, Arad 310130, Romania

ABSTRACT

Tourism is a multi-faced activity that links the economic, social and environmental components of sustainability. This research analyzes rural residents' perceptions of the impact of tourism development and examines the factors that influence the support for sustainable tourism development in the region of Nord-Vest in Romania. Residents' perceptions towards tourism development were measured using 22 items, while their support for tourism development was determined using 8 items. Descriptive and inferential statistics were used to analyze the data. Principal component analysis grouped the first 22 variables into 4 factors, and the following 8 variables into 2 factors (sustainable development, destination development). Findings indicate that residents see tourism as a development factor. The natural, economic, and social-cultural environment as well as infrastructure, age, gender and education are factors that influence the sustainable development of tourism.

Keywords: rural tourism; tourism planning; residents' attitudes; tourism impact

1. INTRODUCTION

The importance of tourism nowadays is determined by the multiple roles that it plays within any country (economic, social and cultural) and its ability to create a positive impact (employment, wealth, dynamism, income enhancement, infrastructure, international friendship and moving people and assets) [1,2]. The tourism sector has become, during the last several decades, a major factor of importance in world GDP because of its rapid growth and development [3,4], often associated with an export industry [5]. Eshliki and Kaboudi [6] describe the tourism sector as a "powerful force of change in the economy," mainly due to its positive economic impact on communities. Stetic [7] underlines the key role that rural areas play in the tourism sector, not only because of their special position as areas of excellence, but also as ecological oases combined with the ability to preserve traditional culture and ethno-cultural heritage. Therefore, Vazques et al. [8] appreciate the importance of rural tourism as a key factor for proper socio-economic development, while Giannakis [9] highlights the importance of rural areas within the European Union from the point of view of their vast territory held (91%) and the population involved (59%). These rural areas can be strengthened and revitalized only through proper and viable tourism based on sustainable development [10]. Because of the high importance of the tourism sector, the newest Tourism Action Framework at the European level mentions among its main priorities the stimulation of long-term competitiveness and the promotion of sustainable development [11]. The main advantages of the sustainable approach in tourism are related to good practices and improvements, especially regarding the environment; however, at the same time, the concepts' incapacity for a proper implementation in practice has been observed [12]. Either way, sustainable tourism development is strongly related to local communities and their attitudes towards tourism. Within the process of developing a sustainable tourism sector, the local community is the key element, as it is directly affected by its evolution [6].

Development of tourism influences the development of other sectors such as agriculture, food processing and handcrafting, which can contribute to the well-being of the entire community. Previous studies about Romanian rural tourism analyzed the factors that influence rural tourism development [13,14], the particularities of rural tourism potential [15,16]], supply [17,18] and demand [17,19]. Local residents represent an important component of the amalgam that constitutes the destination [20] and influences the future development of any tourism destination.

The aim of this paper is to identify the support of the local community towards sustainable tourism development in the region of Nord-Vest in Romania. To this end, residents' perceptions towards the impact of tourism development are identified, and, secondly, the residents' support for future tourism development is determined. The analysis reveals that the support for future tourism development is based on two components: sustainable development and destination development. In this context, the paper presents an original approach towards the subject, the results of which can be useful in creating future development strategies.

2. LITERATURE REVIEW

The idea of sustainable tourism development emerged in the last several decades as a necessity to ensure an efficient tourism sector based on three main components: environmental interests, socio-cultural and economic needs of the communities involved [21]. To maintain a balance between the positive and negative impacts that the three factors could generate, one must take into consideration the local community—the core element within the tourism development process and the most important stakeholder [22,23,24]. Analyzing the local community's perceptions regarding tourism impact becomes a major concern, because it is strongly connected to the will to support tourism development [25,26]. Generally, tourism is perceived by residents as having strong economic benefits, which outweigh any other possible negative impacts, encouraging residents to perceive tourist activity in a positive way and resulting in strong involvement and support [26,27]. Still, interesting findings were noted in Vietnam, where touristic activity is supported by the community not for its economic benefits, but rather for its socio-cultural and environmental benefits [28].

Studies focus both on the negative and positive impacts of tourism on local communities. For example, communities in Egypt have been shown to support tourist activity, even if both negative and positive impacts are perceived, because the positive are stronger than the negative [3]. In terms of economic development, positive impacts are reflected in job creation, investments and increasing the national income, while negative impacts are related to an increased level of inflation. Socio-cultural factors can generate two kinds of attitudes: positive ones related to cultural exchange, and negative ones linked to the way of life and overcrowded places [3]. In research conducted in Arizona, Mcgehee and Andereck [27] concluded that support for tourism development is highly related to personal benefits obtained by the local community members, where overall the community perceives tourism in a positive way due to its effects on the local economy (such as job creation and an improvement in quality of life). A study conducted in Mauritius observed a similar positive economic impact due to job creation and improvements in quality of life. A positive relationship was observed between the support for tourism and economic and socio-cultural factors, while a negative relationship was established between community support and the environment [5].

It has been observed that the main components of sustainable tourism development influence the local community's goodwill and support for tourism; therefore, much research focuses on studying its economic, environmental and socio-cultural impacts [3,5,22,27,29,30]. Even if tourism offers important positive benefits, its strong development could cause major dissatisfaction among communities due to intensive traffic, inflation, and crime [6,29].

The relationship between community satisfaction and tourism development analyzed by Min et al. (2012) [22] shows that, even if the level of satisfaction related to tourist activity in the area is not high, the community still supports it when it is considered promising for future city development. The residents' support for tourist activity is connected also to their perceptions of economic, environmental and socio-cultural factors. Positive impact on aforementioned factors indicate high support from the local community [23,30,31].

An important finding was obtained by Koa and Stewart [29] after conducting a study in the Keju Islands of Korea, a relatively undeveloped area. Results indicated that residents' attitudes toward tourism are directly related to the stage of development of the host community. In Uganda it was observed that the local community has a positive attitude towards tourism because it is considered as a factor of development generating incomes, increasing agricultural production and "good fortune" [32]. The results of research conducted in North Carolina indicated two main reasons for which the community is willing to support tourist activities, which are strongly related to the personal benefits obtained: The female population positively perceives the cultural dimension represented by the development of arts, craft and household items, while the youngsters perceive improvements in social life and recreation facilities [33]. The community attachment and involvement, as well as the support for sustainable tourism development, are influenced by their perceived benefits to local residents [34]. Understanding the residents' perceptions towards tourism is important in order to shape future policies that minimize the potential negative impact of tourism and maximize its benefits [35].

3. MATERIALS AND METHODS

The study was conducted in a rural area in the region of Nord-Vest in Romania from November 2014 to April 2015. Nord-Vest comprises 6 counties: Bihor, Bistrita-Nasaud, Cluj, Maramures, Satu-Mare and Salaj. It has a surface area of 34,159 km2, representing 14.32% of the total country surface. The region borders both Hungary and Ukraine [36].

This region has a high potential for tourism due to its natural environment and landscape (mountains, natural reservations, thermal waters, salt mines) and to a variety of cultural tourist attractions (churches, wooden churches, traditions, etc.). The region includes 170 protected areas of national importance, of which two are national parks (Rodna Mountains National Park and Calimani National Park) and two natural parks (Apuseni Natural Park and Maramures Mountains Natural Park) [37]. The parks, both national and natural, attract tourists due to the possibilities of hiking, bird watching, rural tourism and agritourism [38,39].

Table 1 presents a timeline of tourism's supply and demand from 2005 to 2014. In Nord-Vest, at the end of 2014, there were 676 (11% of the total number at national level) lodging facilities, of which 33% (226) were agritourism guesthouses, which represent the main type of accommodation facility in the rural area. During the last 10 years in Nord-Vest, the number of lodgings rose by 40% and the number of agritourism guesthouses 61%, while the occupancy rate and the average length of stay remained quite similar. This can be explained by the increase in the number of tourist arrivals and overnight stays. The arrivals in agritourism guesthouses in the region of Nord-Vest increased by almost 200%, and the overnight stays by 188% from 2005 to 2014. The number of guesthouses increased from 2005 to 2010 by 90% mainly due to European funds that supported rural development, and by the end of 2014 they decreased by 15%. Factors which could have led to this decreasing trend may include managers' lack of specific entrepreneurial education and skills [40], and owners' lack of capacity to adapt and diversify their services according to the needs of the tourists [41]. The majority of the guesthouses assure basic facilities (bed & breakfast); in this case, it can be stated that Romanian rural tourism lacks additional touristic services for entertainment, outdoor and indoor recreation, handicrafts and other souvenirs [42].

Table 1. Evolution of tourist activity indicators in Romania and North-West Region.

Region	Indicator	Year									
		2005	2006	2007	2008	2009	2010	2011	2012	2013	2014
Romania	Number of lodgings	4226	4710	4694	4840	5095	5222	5003	5821	6009	6130
	Number guest houses	956	1259	1292	1348	1412	1354	1210	1569	1598	1665
	Total average length of stay (nights) *	3.16	3.06	2.95	2.91	2.82	2.64	2.56	2.49	2.44	2.40
	Guesthouses average length of stay (nights) *	2.15	2.12	2.05	2.08	2.07	2.09	2.06	2.03	1.99	1.97
	Total occupancy rate (%) *	33.42	33.61	36.04	35.02	28.35	25.16	26.28	25.85	25.14	26.11
	Guest houses occupancy rate (%) *	14.47	14.41	16.34	18.41	14.22	12.36	13.78	13.20	12.56	13.16
Northwest	Number of lodgings	480	543	554	585	645	658	650	730	709	676
	Number guest houses	140	193	200	225	259	266	206	252	231	226
	Total average length of stay (nights) *	3.12	3.03	2.87	2.79	2.87	2.68	2.61	2.48	2.35	2.35
	Guesthouses average length of stay (nights) *	1.99	1.98	1.92	1.95	1.93	1.98	2.04	2.03	1.92	1.92
	Total occupancy rate (%)*	32.24	32.06	34.05	32.69	27.70	23.25	23.61	22.33	22.58	25.38
	Guest houses occupancy rate (%) *	13.16	15.17	19.74	24.22	14.52	9.35	9.77	10.76	10.95	13.15

Source: National Institute of Statistics, Tempo-online time-series [43]; * compute based on data from National Institute of Statistics.

To examine residents' perceptions towards the impact of tourism development in Nord-Vest, the authors used a quantitative survey. The collected data can be divided into three main sections: perception of tourism impact, support of tourism development and socio-demographic characteristics of the respondents. The 22 variables used to determine the rural residents' perceptions of tourism development were adopted from previous studies [3,5,6,27,29,33,44,45,46,47]. The support for tourism development was measured with 8 variables developed by [47,48]. A 5-point Likert-type scale was used based on the following scale: 1 = strongly disagree, 2 = disagree, 3 = neutral, 4 = agree, 5 = strongly agree to evaluate each variable.

Data were collected from 433 residents from the rural area of Nord-Vest. The sample size was determined based on the following formula [30].

$$n = \frac{(1.96)^2 \times 1075725 \times 0.5 \times (1-0.5)}{(0.05)^2 \times (1075725 - 1) + (1.96)^2 \times 0.5 \times (1-0.5)} = 400 \text{residents}$$

(1)

550 questionnaires were distributed to a proportional stratified random sample of adults. The response rate was 91% (502 questionnaires), and in the end 433 questionnaires were validated.

The selection of the respondents was based on their age respective of the distribution of the original population.

Descriptive statistics were used to determine the socio-demographic profile of the rural residents and to describe the rural residents' perceptions of tourism development as well as their support for it. Exploratory factor analysis was used to assess the factor structure of the variables. Principal component analysis (PCA) is a data reduction method in which the components are calculated using all of the variance of the manifest variables, with all of that variance appearing in the solution [49]. This is achieved by transforming a new set of variables, which are uncorrelated, and which are ordered so that the first few explain most of the variance [50]. Mathematically, this is equivalent to finding the best low rank approximation of the data via the singular value decomposition [51]. Two principal component analyses were conducted separately, the first to group the perception variables about tourism development, and the second to group the variables regarding future tourism development support. The varimax rotation was used to maximize the differences among the components extracted and to maintain correlation among the components. A simple correlation analysis (Spearman rank-order correlation coefficient) was utilized to calculate the correlation between the support for sustainable tourism development and the impact of tourism, and between the support for tourism destination development and the impact of tourism development.

4. RESULTS AND DISCUSSION

4.1. Rural Residents' Characteristics

The socio-demographic characteristics of the respondents are presented in Table 2. According to the National Institute of Statistics, at the end of the year 2014, 57% of people from rural areas were more than 40 years old. This data is reaffirmed by the results of the current study. The results can mainly be attributed to the lack of attractive employment opportunities in the rural area for the younger generation. Another problem in the rural area is the low level of education. Most of the respondents had graduated from high school (45.4%), while 28% had less than high school education. The monthly household income levels reported are less than 225 Euro (36.4%), 225–445 Euro (36%), more than 445 Euro (27%), while the average family number is 3.7 members. It can be concluded that the rural population was aging and less educated, with a low monthly income.

Table 2. Characteristics of the respondents.

Variables	%
Gender	
Female	41.57
Male	58.43
Age	
15–19	1.20
20–29	18.2
30–39	24.1
40–49	28.8
50–59	19.6
>60	8.00
Education	
Illiterate	0.20
Less than high school	28.1
High school	45.4
University	26.2
House hold income	
<225 euro	36.4
225–445 euro	35.9
>445 euro	27.7

4.2. Factors of Rural Residents' Perception towards Tourism Development

Principal factor analysis was conducted to assess the dimensionality of the 22 items. The Barlett test of sphericity is significant (Chi-square = 3915.62, p < 0.000). The Kaiser-Meyer-Olkin (KMO) overall measure of sampling is 0.88, indicating that data are suitable for the principal component analysis [52]. Values of 0.6 or above from the KMO measures indicated that data are adequate for PCA [53]. The PCA with varimax rotation of the 22 variables resulted in a four-component solution that explains 57.66% of the total variance. Only factors with eigenvalues greater than one were selected [54]. Cronbach's alpha reliability coefficient was computed to evaluate the internal consistency of each component. An acceptable reliability coefficient is higher than 0.6 [54,55,56]. The overall reliability of the 22 variables was 0.87.

The components recorded after the first principal component analysis are listed in Table 3. Component 1 comprises 8 variables (0.87 alpha), component 2 comprises 6 variables (0.82 alpha), component 3 comprises 5 variables (0.73 alpha), and component 4 comprises 3 variables (0.67 alpha).

Table 3. Principal component analysis on tourism impact variables.

Eigenvalue	Variance %	Component	Item	Factor Loading	Communalities
6.10	27.74	Environmental $\alpha = 0.87$	Development of tourism damage natural environment and landscape	0.793	0.632
			Tourism cause overcrowding problems for residents	0.777	0.661
			Tourism increase the air pollution	0.775	0.641
			Tourists use too much water	0.770	0.598
			Tourism results in more litter in an area	0.739	0.571
			Tourism development negatively affects the recreational facilities and entertainment	0.690	0.537
			The construction of tourist facilities destroy the environment	0.674	0.461
			Increase traffic problems	0.561	0.490
3.67	16.68	Economical $\alpha = 0.82$	Tourism plays an important role in the economic development of the area	0.824	0.696
			Tourism improves locals standard of living	0.802	0.735
			Tourism increases a community's tax revenue	0.699	0.557
			Tourism create new jobs for locals	0.649	0.562
			Tourism diversifies the rural economy	0.546	0.502
			Tourism results in an increase in the cost of living	0.540	0.624
			Revenue from tourism taxes activity should be invested in future development of tourism	0.523	0.386
1.88	8.56	Social and Cultural $\alpha = 0.73$	Tourism provide incentives for restoration of traditional houses	0.781	0.639
			Interaction with tourists is a positive experience	0.776	0.617
			Shopping and restaurants option is better as a result of tourism	0.676	0.552
			Tourism development enhance more recreational opportunities for locals	0.431	0.468
1.03	4.67	Infrastructure $\alpha = 0.67$	Improves traffic network	0.798	0.693
			Improves living utilities infrastructure (supply of water. sewage. electric etc.)	0.711	0.583
			Quality of public services in better	0.508	0.480
Total variance %	57.66				

Table 4. Perception on tourism development impact.

Item	Mean	SD
Environment	2.58	1.049
Damage natural environment and landscape	2.45	1.393
Tourism cause overcrowding problems for residents	2.61	1.487
Tourism increase the air pollution	2.52	1.446
Tourists use too much water	2.33	1.398
Tourism results in more litter in an area	3.02	1.490
Tourism development negatively affects the recreational facilities and entertainment	2.60	1.463
The construction of tourist facilities destroy the environment	2.47	1.360
Increase traffic problems	2.69	1.412
Economic	3.40	0.913
Tourism plays an important role in the economic development of the area	3.78	1.235
Tourism improves locals standard of living	3.31	1.290
Tourism increases a community's tax revenue	3.73	1.294
Tourism create new jobs for locals	3.07	1.435
Tourism results in an increase in the cost of living	2.92	1.391
Tourism diversifies the rural economy	2.97	1.323
Revenue from tourism taxes activity should be invested in future development of tourism	4.09	1.158
Social and Cultural	3.5	1.008
Tourism provide incentives for restoration of traditional houses	3.35	1.488
Interaction with tourists is a positive experience	3.58	1.321
Shopping and restaurants option is better as a result of tourism	3.44	1.391
Tourism development enhance more recreational opportunities for locals	3.62	1.197
Physical	3.19	1.100
Improves traffic network	3.06	1.508
Improves living utilities infrastructure (supply of water, sewage, electric *etc.*)	3.19	1.459
Quality of public services in better	3.33	1.261

The first component labeled "environmental effects" explains 27.74% of the variance and has a mean of 2.58 (SD = 1.049). This component involves attributes that focus on conservation of natural resources and negative impacts of tourism on the environment. The environmental effects are seen as being the most negative for the rural residents in terms of tourism development, such as destroying the natural environment (factor loading 0.793), overcrowding problems (factor loading 0.777), air pollution (factor loading 0.775), and water scarcity (0.770). The second component labeled "economic benefits" explains 16.68% of the variance and has a mean of 3.40 (SD = 0.913). This component involves attributes related to the overall economic development (factor loading 0.824), tax revenue (factor loading 0.699), employment (factor loading 0.649), and living cost (factor loading 0.540). The third component labeled "social and cultural impacts" involves attributes related to quality of life and cultural activities. It explains 8.56% of the variance and has a mean of 3.50 (SD = 1.008). This component groups items related to restoration of traditional houses (factor loading 0.781), diversification of recreational facilities (factor loading 0.431), alternative possibilities for shopping and dining (factor loading 0.676), and the perception of residents' interaction with tourists (factor loading 0.776). The fourth component explains 4.67% of the variance and involves attributes related to infrastructure development, being labeled "infrastructure benefits"; the mean is 3.19 (SD = 1.100) (Table 3 and Table 4). These results diverge from previous research, which has shown either that economic impacts may be more important [46,47], or that the social and cultural impacts were more significant [5] than the environmental component.

4.3. Local Residents' Perception towards Tourism Development Impacts

The host community tends to agree that development of tourism has a positive impact on the development of the region. The effects of tourism development on the natural environment and landscape are not perceived as being negative. The mean value of component 1 is 2.58, lower than the one found in a similar study [5]. The only variable with a higher score than 3 is the one related to the quantity of litter in the area.

The rural residents believe that the revenue from the taxation of touristic activity should be used for the future development of this sector (mean = 4.09), since an increase in the community's tax revenue is perceived (mean = 3.73). Tourism is seen as an important factor for the economic development of the area (mean = 3.78), which improves local standard of living (mean = 3.31) without affecting the cost of living (mean = 2.92). The findings support past studies regarding the people's perception towards the economic impact of tourism [27,46,47]. At the same time, it should be underlined that, for the rural residents of Nord-Vest, tourism is not perceived as an alternative to agricultural activities, nor as a factor of economic activities' diversification (mean = 2.97). One of the causes may be the lack of knowledge regarding the founding sources and the fear of business failure [40].

The rural residents agree that the development of tourism in their region provides more recreational opportunities (mean = 3.62), and interaction with the tourists is perceived as a positive experience (mean = 3.58). The rural residents perceive the development of tourism as a factor that provides cultural identity and improves the quality of services, and, indirectly, the standard of living. These results confirm the findings of previous studies [46,47]. Tourism leads to the restoration and preservation of the cultural values of the rural community (mean = 3.35). This is an important aspect for assuring sustainable development of the rural area.

Tourism is perceived as a factor that influences the development of the traffic network (mean = 3.06) and of the living utilities (mean = 3.19) and improves the quality of public services (mean = 3.33). Improvement in the quality of public services is a result of higher standards of the tourists who visit the area. Therefore, in order to satisfy the tourists' expectations, the services are adapted to their needs and quality standards are imposed.

4.4. Local Residents' Support for Future Tourism Development

Principal component analysis was used to assess the reliability of the 8 variables related to future sustainable tourism development. The Barlett test of sphericity is significant (Chi-square = 1026.348, $p < 0.000$). The Kaiser-Meyer-Olkin (KMO) overall measure of sampling is 0.84, indicating that data are suitable for the principal component analysis [52]. The PCA with varimax rotation of the 8 variables resulted in a two-component solution that explains 56.67% of the total variance. The overall reliability of the 8 variables is 0.83.

Four attitude variables ("plans are important to manage the growth of tourism," "long-term planning reduces the negative environmental impact," "authorities support tourism development," and "new environment protection measures") concerning the sustainable development of tourism were loaded in the first component with the cross-correlation coefficients of 0.820, 0.751, 0.747 and 0.703. This factor accounts for 42.89% of the total variance and was named support for tourism sustainable development (Table 5). The higher scores indicate that rural residents are focused on the long-term impact of tourism development.

The second component consists of 4 variables ("tourism represents a sustainable activity," "support for new facilities," "my community should become a tourist destination" and "tourism should become an important part of the community") with the cross-correlation coefficients of 0.756, 0.740, 0.665 and 0.514 (Table 5). This factor accounts 13.78% of the total variance and was termed support for tourism destination development.

Table 5. Support for future tourism development.

Eigenvalue	Variance %	Component	Items	Factor Loading	Communalities
			It is important to develop plans to manage the growth of tourism	0.820	0.717
3.43	42.89	Sustainable development $\alpha = 0.77$	Long-term planning will reduce the negative environmental impacts	0.751	0.565
			I agree that local authorities support tourism development	0.747	0.625
			New environment protection measures should be developed	0.703	0.544
			Tourism is a sustainable activity in my community	0.756	0.580
1.10	13.78	Tourism destination $\alpha = 0.61$	I support new tourism facilities	0.740	0.617
			My community should become more of a tourist destination	0.665	0.617
			I support tourism and I would like to see it become an important part of my community	0.514	0.270
Total variance %	56.67				

The development of tourism should be supported by the local authorities (mean = 4.35) by setting strategic plans, with clear actions for tourism's growth management (mean = 4.11). The long-term planning with reduction of negative environmental impacts (mean = 3.76) and new environmental protection measures (mean = 3.95) are less supported than the involvement of local authorities and strategic planning of tourism development because of the reduced negative impact of tourism on the environment, as was noticed in Section 4.3.

The rural residents believe that tourism should be encouraged and become an important part of the community (mean = 4.12). This can be achieved by developing new tourism facilities (mean = 4.32) that will assure job alternatives for the local community. Furthermore, development plans should take into consideration the desire of the local residents for the sustainable development of their region as a tourism destination (mean = 4.12) (Table 6).

Table 6. Degree of agreement for future tourism development.

Statements	Mean	SD
Sustainable development	4.03	0.961
It is important to develop plans to manage the growth of tourism	4.11	1.196
Long-term planning will reduce the negative environmental impacts	3.76	1.273
I agree that local authorities support tourism development	4.35	1.111
New environment protection measures should be developed	3.95	1.331
Destination development	4.24	0.820
I support new tourism facilities	4.32	0.994
Tourism is a sustainable activity in my community	4.12	1.134
My community should become more of a tourist destination	4.12	1.118
I support tourism and I would like to see it become an important part of my community	4.49	1.678

The results of simple correlation analysis on the support for sustainable tourism development and support for destination development for environmental impacts, economic benefits, socio-cultural impacts, infrastructure benefits, and age are listed in Table 7. The environmental impact ($r = -0.214$, $p < 0.01$), economic benefits ($r = 0.230$, $p < 0.01$), socio-cultural impacts ($r = 0.498$, $p < 0.01$) and infrastructure benefits ($r = 0.328$, $p < 0.01$) were significantly correlated with support for sustainable development. The environmental impact ($r = -0.252$, $p < 0.01$), economic benefits ($r = 0.241$, $p < 0.01$), socio-cultural impacts ($r = 0.418$, $p < 0.01$), and infrastructure benefits ($r = 0.292$, $p < 0.01$) were significantly correlated with support for future tourism destination development. Age does not have any influence on the residents' support for tourism development (Table 7). A t-Test was carried out to examine the influence of gender and education level on the support for future tourism development (Table 8 and Table 9).

Table 7. Correlation of each variable with the support for sustainable development and destination development.

Dependent Variable	Sustainable Development		Destination Development	
Independent Variable	Correlation Coefficient (r)	p-Value	Correlation Coefficient (r)	p-Value
Environment impact	−0.214 **	0.000	−0.252 **	0.000
Economic benefits	0.230 **	0.000	0.241 **	0.000
Social and cultural impacts	0.498 **	0.000	0.418 **	0.000
Infrastructure	0.328 **	0.000	0.292 **	0.000
Age	−0.036 *	0.457	−0.052	0.282

* significant at 0.05; ** significant at 0.01.

Table 8. Results of t-test analysis of gender and support for future tourism development.

Dependent Variable	Means		t Value	Sig.
	Female	Male		
Sustainable development	4.17	3.94	2.545	0.013 *
Destination development	4.34	4.18	2.017	0.044 *

* significant at 0.05; ** significant at 0.01.

Table 9. Results of t-test analysis of education level and support for future tourism development.

Dependent Variable	Means		t Value	Sig.
	Less than High School	More than High School		
Sustainable development	3.91	4.08	−1.696	0.091
Destination development	4.14	4.29	−1.715	0.087

* significant at 0.05; ** significant at 0.01.

There is an indirect link between perceived environmental impact and the support for sustainable development, and between environmental impact and support for tourism destination development. A direct link can be observed between economic impact, social and cultural impact, infrastructure benefits and the support of local community for future tourism development (Table 7).

Residents who perceive the impact of tourism less negatively tend to more highly support sustainable tourism development in the rural area. Referring to Table 8, the results show that females differ significantly in their support for tourism development (mean = 4.17 for sustainable development and mean = 4.34 for destination development). This can be explained by the fact that tourism represents an alternative activity for agriculture in the rural space, which can attract young people to establish themselves and work in the rural area. Furthermore, tourism represents a source of employment and entrepreneurial opportunities for women [57]. Results showed that residents who perceive the environmental impact of tourism more negatively tend to display diminished support for the future development of tourism, as was also suggested in previous studies [5,44].

Economic and socio-cultural benefits can be considered personal benefits of tourism development. The greater the benefits the rural residents perceive from tourism, the more likely they are to support sustainable tourism development, and the more likely they are to transform their community into a tourism destination. The findings support past studies regarding residents' support of tourism development, which found that the higher the personal benefits from tourism are, the more willing the local residents are to develop tourism in their community [28,33].

5. CONCLUSIONS

The research objectives were to investigate rural residents' attitudes towards various tourism impact variables and to explore their influence on the support of future sustainable tourism development. The results of this research indicate that rural residents perceive tourism development positively. Tourism is a sustainable development activity in the rural community. The environmental component of sustainable development is the most important one, a fact that can be explained by the high natural tourist potential and the awareness of the local residents of the importance of natural conservation for sustainable development, on the one hand, and the desire to reduce the negative effects of tourism development on the area on the other.

Tourism development improves the quality of life of local residents due to its effect on economic development of the area, which in turn leads to new employment opportunities. Furthermore, tourist activity in the rural area is perceived as being beneficial to the diversification of recreational alternatives and the improvement of the general infrastructure. Similar findings in examination of residents' attitudes toward tourism development were observed by Abdollahzadeh and Sharifzadeh [46].

Results indicate that tourism impacts are perceived positively as employment opportunity and well-being increases. An important segment of the rural population is willing to support the development of sustainable tourism because of the personal benefits obtained in terms of socio-cultural aspects (arts and crafts development, improvement of social life and facilities).

The results show that rural residents see tourism as an income generator, but at the same time they understand the importance of planning and managing tourism destinations sustainably. The local community, as an important stakeholder within the tourism sector, becomes a key element in developing future tourism strategies. Hanafiah [47] underlined that the community should get actively involved in the process of tourism development.

For rural residents, tourism is seen as an opportunity to enhance the wellbeing of the community in general, and particularly their own. The local community is willing to support sustainable tourism development if the personal benefits perceived are important. The greater the perceived economical, socio-cultural and infrastructure benefits, the higher the support is from the local community in building future tourism strategies.

This study revealed a lack of knowledge regarding residents' perceptions towards tourism impact and attitudes toward sustainable tourism development in a rural area of Romania. The results provide tourism planners and policymakers with viable information regarding rural residents' attitudes toward the development of tourism, which can serve as a useful tool in future development plans, enhancing sustainable tourism development and reducing tourism's negative impacts. Even if at the moment the residents tend to support sustainable tourism development, it is recommended that long-term changes in residents' attitudes and perceptions be taken into account. Furthermore, it was shown that the support for sustainable tourism development is greater if policymakers and tourism planners are attentive to residents' concerns and beliefs [28].

Over time, a lack of consistent and reliable information regarding local residents' attitudes towards sustainable rural development has negatively influenced the decision-making process regarding the sector's funding allocation. In fact, The National Rural Development Plan 2014–2020 of Romania includes two different specific measures indirectly related to rural tourism: Measure 06—"Development of exploitations and companies" (Under-measure 6.4 "Support for investments in creating and developing non-agricultural activities"), and Measure 07—"Basic services and village renewal in rural areas" (Under-measure 7.6 "Investments related to cultural heritage protection") [58]. No specific measure was dedicated to rural tourism during this period. In the first two years of implementation (2014 and 2015), 52 proposals were submitted (75,987,178 Euro) under Under-measure 6.4, while only 2 were accepted and financed (350,858 Euro), and none were financed in the first two years. Concerning Under-measure 7.6, 224 proposals were submitted (42,565,798 Euro), and not one was accepted and financed until the end of 2015 [59]. The results of this research can contribute to an improved strategy regarding rural tourism development, such as considering residents' requirements and expectations when creating the specifications of the under-measures, a process that would result in more sustainable rural tourism development.

Finally, this study has several limitations: primarily, limited time and low budget. Like its predecessors [27], this study did not clarify how the residents perceive themselves as benefiting from sustainable tourism development. Due to the variety of natural and anthropic tourism potential at this moment, based on the results of the research, it is impossible to formulate specific recommendations for different types of tourism products. Furthermore, the studies that follow should focus more on the particularities of the natural characteristics of the research area (thermal water, natural parks proximity, etc.), while more factors influencing the support for sustainable tourism development should be added (field of activity, land ownership, dependence on tourism, preoccupation for nature protection and responsible behavior) to determine the support for sustainable tourism development. In this way, important information can be gained and geared towards specific actions for future tourism development.

ACKNOWLEDGMENTS

Assistance provided by Molly McDonough, graduate of Master Food Identity, during the writing and English-proofing of the manuscript is greatly appreciated.

AUTHOR CONTRIBUTIONS

Iulia C. Muresan, Rezhen Harun and Felix H. Arion designed the research; Camelia F. Oroian, Alexandru Todea and Ramona Lile applied the survey; Camelia F. Oroian and Andra Porutiu organized the data based; Iulia C. Muresan and Gabriela O. Chiciudean analyzed the data, Iulia C. Muresan, Gabriela O. Chiciudean, Felix H. Arion and Rezhen Harun wrote the manuscript.

REFERENCES

1. Egbali, N.; Nosrat, A.B.; Alipour, S.K.S. Effects of positive and negative rural tourism (Case study: Rural Semnan Province). J. Geogr. Reg. Plan. 2011, 4, 63–76.

2. Shariff, N.M.; Abidin, A.Z. Community attitude towards tourism impacts: Developing a standard instrument in the Malaysian context. E-J. Soc. Sci. Res. 2013, 1, 386–396.

3. Eraqi, M.I. Local communities' attitudes towards impacts of tourism development in Egypt. Tour. Anal. 2007, 12, 191–200.

4. Budeanu, A. Impacts and responsibilities for sustainable tourism: A tour operator's perspective. J. Clean. Product. 2005, 13, 89–97.

5. Ramseook-Munhurrun, P.; Naidoo, P. Residents' attitudes toward perceived tourism benefits. Int. J. Manag. Mark. Res. 2011, 4, 45–56.

6. Eshliki, S.A.; Kaboudi, M. Perception of community in tourism impacts and their participation in tourism planning: Ramsar, Iran. J. Asian Behav. Stud. 2012, 2, 51–64.

7. Stetic, S. Specific features of rural tourism destinations management. J. Settl. Spat. Plan. 2012, 1, 131–137.

8. De la Torre, G.M.V.; Gutiérrez, E.M.A.; Guzman, T.J.L.G. Tourism as generator of wealth in rural areas. In Proceedings of the 18th European Advanced Studies Institute in Regional Science, Lodz-Cracow, Poland, 1–10 July 2005.

9. Giannakis, E. The role of rural tourism on the development of rural areas: The case of Cyprus Elias. Romanian J. Reg. Sci. 2014, 8, 38–53.

10. Garau, C. Perspectives on cultural and sustainable rural tourism in a smart region: The case study of Marmilla in Sardinia (Italy). Sustainability 2015, 7, 6412–6434.

11. Iunius, R.F.; Cismaru, L.D. Raising competitiveness for tourist destinations through information technologies within the newest tourism action framework proposed by the European commission. Sustainability 2015, 7, 12891–12909.

12. Mihalic, T. Sustainable-responsible tourism discourse e towards 'responsustable' tourism sustainability. J. Clean. Prod. 2014.

13. Turnock, D. Sustainable rural tourism in the Romanian Charpatians. Geogr. J. 1999, 165, 192–199.

14. Dragulanescu, I.V.; Drutu (Ivan), M. Rural tourism for local economic development. Int. J. Acad. Res. Account. Financ. Manag. Sci. 2012, 2, 196–203.

15. Iorio, M.; Corsale, A. Rural tourism and livelihood strategies in Romania. J. Rural Stud. 2010, 26, 152–162.

16. Negrusa, A.L.; Cosma, S.A.; Bota, M. Romanian rural tourism development: A study case in Maramures. Int. J. Bus. Res. 2007, 7, 129–135.

17. Marin, D.; Petroman, C.; Petroman, I.; Ciolac, R.; Balan, I. Study regarding rural guest-house and agritourist household's number and percent in the total number of tourists establishments in Romania. Lucr. Stiintifice Ser. Agron. 2009, 52, 469–470.

18. Arion, F.; Muresan, I. An overview of rural tourism in the development regions of Romania. Bull. UASVM Hortic. 2008, 62, 35–40.

19. Naghiu, A.; Vasquez, J.L.; Georgiev, I. Rural development strategies through rural tourism activities in Romania: Chance for an international demand? Int. Rev. Public Nonprofit Mark. 2005, 2, 85–95.

20. Vanhove, N. The Economics of Tourism Destination; Elsevier Butterworth-Heinemann: Oxford, UK, 2005.

21. Mansfeld, Y.; Jonas, A. Evaluating the socio-cultural carrying capacity of rural tourism communities: A 'value stretch' approach. Tijdschr. Voor Econ. Soc. Geogr. 2006, 97, 583–601.

22. Min, Z.; Xiaoli, P.; Bihu, W. Research on residents' perceptions on tourism impacts and attitudes: A case study of Pingyao ancient city. In Proceedings of the 6th Conference of the International Forum on Urbanism (IFoU), Tourbanism, Barcelona, 25–27 January 2012; pp. 1–10.

23. Mohammadi, M.; Khalifah, Z.; Hosseini, H. Local people perceptions toward social, economic and environmental impacts of tourism in Kermanshah (Iran). Asian Soc. Sci. 2010, 6, 220–225.

24. Banki, M.B.; Ismail, H.N. Multi-stakeholder perception of tourism impacts and ways tourism should be sustainably developed in obudu mountain resort. Dev. Ctry. Stud. 2014, 4, 37–48.

25. Huh, C.; Vogt, C.A. Changes in residents' attitudes toward tourism over time: A cohort analytical approach. J. Travel Res. 2008, 46, 446–455.

26.   Bestard, A.B.; Nadal, R.J. Attitudes toward tourism and tourism congestion. Reg. Dev. 2007, 25, 193–207.
27.   McGehee, N.G.; Andereck, K.L. Factors predicting rural residents' support of tourism. J. Travel Res. 2004, 43, 131–140.
28.   Long, P.H. Perceptions of tourism impact and tourism development among residents of Cuc Phuong National Park, Ninh Binh, Vietnam. J. Ritsumeikan Soc. Sci. Hum. 2011, 3, 75–92.
29.   Ko, D.W.; Stewart, W.P. A structural equation model of residents' attitudes for tourism development. Tour. Manag. 2002, 23, 521–530.
30.   Brida, J.G.; Disegna, M.; Osti, L. Residents' perceptions of tourism impacts and attitudes towards tourism policies in a small mountain community. In Proceedings of the Ninth Canadian Congress on Leisure Research, Wolfville, NS, Canada, 12–15 May 1999.
31.   Choi, H.C.; Murray, I. Resident attitudes toward sustainable community tourism. J. Sustain. Tour. 2010, 18, 575–594.
32.   Lepp, A. Residents' attitudes towards tourism in Bigodi village, Uganda Case study. Tour. Manag. 2007, 28, 876–885.
33.   Wang, Y.; Pfister, R.E. Residents attitudes toward tourism and perceived personal benefits in a rural community. J. Travel Res. 2008, 47, 84–93.
34.   Lee, T.H. Influence analysis of community resident support for sustainable tourism development. Tour. Manag. 2013, 34, 37–46.
35.   Stylidis, D.; Biran, A.; Sit, J.; Szivas, E.M. Residents' support for tourism development: The role of residents' place image and perceived tourism impacts. Tour. Manag. 2014, 45, 260–274.
36.   Agenţia de Dezvoltare Regională Nord-Vest. Prezentarea Regiunii Nord-Vest (Transilvania de Nord). Available online: http://www.nord-vest.ro/DESPRE-NOIAgentia-de-Dezvoltare-Regionala-Nord-Vest/REGIUNEA-TRANSILVANIA-DE-NORD/Prezentare-Regiune.html (accessed on 1 November 2015). (In Romanian).
37.   The Convention on Biological Diversity. Regiunea Nord-Vest (Cluj-Napoca). Available online: http://biodiversitate.mmediu.ro/romanian-biodiversity/despre-arii-protejate/arpm/regiunea-nord-vest-cluj-napoca/ (accessed on 1 November 2015). (In Romanian).
38.   Dumitraş, D.; Pop, A. Perspective on the management of Rodna Mountains National Park. Bull. UASVM Hortic. 2009, 66, 164–169.
39.   Dumitras, D.E.; Muresan, I.C.; Ilea, M.; Jitea, I.M. Agritourism—A potential linkage between local communities and parks to maintain sustainability. Bull. UASVM Hortic. 2013, 70, 300–309.
40.   EY, Building a Better Working World. Entrepreneurs Speak Out. Entrepreneurship Barometer Romania 2013. Available online:

http://www.ey.com/Publication/vwLUAssets/EY_Entrepreneurs_Speak
_Out_survey_13_Feb_2014_(EN)/$FILE/EY%20Entrepreneurship%20
Barometer%20Romania%202013.pdf (accessed on 30 December
2015).

41. Muresan, I.C.; Arion, F.H.; Harun, R. Study regarding rural guesthouse
and tourists' satisfaction. Bull. UASMV Hortic. 2013, 70, 362–367.

42. Ioan, I.; Radulescu, C.V.; Bran, F. Romanian rural tourism: Status and
prospects by innovative organizational approaches. J. Tour. Stud. Res.
Tour. 2014, 17, 15–21.

43. National Institute of Statistics, Tempo-Online Time Series. Available
online: http://statistici.insse.ro/shop/?lang=en (accessed on 25 October
2015).

44. Teye, V.; Sonmez, S.F.; Sirakaya, E. Residents' attitudes toward
tourism development. Ann. Tour. Res. 2002, 29, 668–688.

45. Brida, J.G.; Disenga, M.; Osti, L. Residents' perception and attitudes
towards tourism impacts: A case study of the small rural community of
Folgaria (Trentino—Italy). Int. J. 2011, 18, 359–385.

46. Abdollahzadeh, G.; Sharifzaden, A. Rural residents' perception toward
tourism development: A study from Iran. Int. J. Tourism Res. 2014, 16,
126–136.

47. Hanafiah, M.H.; Jamaluddin, M.R.; Zulkifly, M.I. Local community
attitude and support towards tourism development in Tioman Island,
Malaysia. Procedia Soc. Behav. Sci. 2013, 105, 792–800.

48. Wang, Y.; Pfister, R.E.; Duarte, B. Residents' attitudes toward tourism
development: A case study of Washington, NC. In Proceedings of the
North-Eastern Recreation Research Symposium, Bolton Landing, NY,
USA, 9–11 April 2006; pp. 411–418.

49. Costello, A.B.; Osborne, J.W. Best practices in exploratory factor
analysis: Four recommendations for getting the most from your
analysis. Pract. Assess. Res. Eval. 2005, 10, 1–9.

50. Jollitte, I.T. Principal Component Analysis; Springer Science +
Business Media: New York, NY, USA, 2014.

51. Ding, C.; He, X. K-means clustering via principal component analysis.
In Proceedings of the 21st International Conference on Machine
Learning, Banff, AB, Canada, 4–8 July 2004; pp. 29–37.

52. Kaiser, H.F. Index of factorial simplicity. Psychometrika 1974, 39, 31–
36.

53. Tabachinick, B.G.; Fidell, L.S. Using Multivariate Statistics, 2nd ed.;
Harper & Row: Cambridge, UK, 1989.

54. Hair, J.F.; Anderson, R.E.; Tatham, R.L.; Black, W.C. Multivariate Data Analysis, 5th ed.; Prentice Hall: Upper Saddle River, NJ, USA, 1998.]

55. Burgess, S.M.; Steenkamp, J.B.E.M. Marketing renaissance: How to research in emerging markets advances marketing science and practice. Int. J. Res. Mark. 2006, 23, 337–356.

56. Nunnally, J.; Bernstein, I. Psychometric Theory, 3rd ed.; McGraw-Hill: New York, NY, USA, 1994.

57. Figueroa-Domecq, C.; Pritchard, A.; Segovia-Perez, M.; Morgan, N.; Villace-Molinero, T. Tourism gender research: A critical accounting. Ann. Tour. Res. 2015, 52, 87–103.

58. Ministry of Agriculture and Rural Development in Romania. Programul Național de Dezvoltare Rurală pentru perioada 2014–2020. Available online: http://www.madr.ro/docs/dezvoltare-rurala/programare-2014-2020/PNDR_2014_-_2020_01.07.2014.pdf (accessed on 22 December 2015). (In Romanian).

59. Ministry of Agriculture and Rural Development in Romania. Situatia proiectelor depuse in cadrul PNDR 2014-2020 la data de 31.12.2015 conform registului electronic al ceririlor de finanțare. Available online: http://www.madr.ro/pndr-2014-2020/implementare-pndr-2014-2020/situatia -proiectelor-depuse- 2014-2020/download/1816 _7595b8fd 510ac8d1f2b3cfd18932f357.html (accessed on 19 January 2016). (In Romanian).

CHAPTER 8

An Exploration on the Effective Factors of Tourism Industry on Protection of the Environment in the Historical City Ghoumas

Hooman Mesgarian[1*], Leili Alaei[2]

[1]*Department of Geography, Islamic Azad University, Semnan Branch, Member of Young Researchers Elite Club, Semnan, Iran*

[2]*Department of Architecture, Islamic Azad University, Savadkooh Branch, Savadkooh, Iran*

ABSTRACT

According to the world tourism organization (WHO) in 2005, the number of tourists has increased up to 36 times. In a way that it has maximized from 25 million in 1950 to 1018 million in 2010. The sustainable tourism, thus, because of making a balance between different environmental, economic, cultural and social dimensions of the tourism development, plays a considerable role in protection of species diversity, and benefits the tourism activists in order to reduce its destructive impact on the environment and the local cultures and being kept for the future generations. Accordingly, the present research first examines the relationship between tourism, sustainable development and the ecology, for measurement of different aspects of the industry effect on the environment. Secondly, the present study uses from some parameters like the average monthly minimum and maximum temperatures, the average relative humidity, the average maximum and minimum temperatures in day and at night in the Semnan station as well as using the common diagram in the experimental methods like Olegi to assess day and night Climatourism in the region. Finally, on the basis of the performed analysis, the strengths and weak points of tourist attractions for the protection of the environment are considered.

Keywords: Climatourism, Semnan, Sustainable Development, The Environment, Tourism

1. INTRODUCTION

The tourism industry as one of leading industries including hotel, restaurants, the domestic and international transportation and handcrafts has played a considerable role. According to the data reported by World Tourism Organizations (WTO), the tourism industry in 2005 cost 6.2 trillion dollars [1]. In the world today, tourism has been recognized as a factor for improvement of quality of life in the developing countries. Sustainable development, in fact, due to its ability in making a balance between ecological economic and cultural dimensions of tourism development is highly effective in protection of species diversity. The tourism industry struggles to create jobs, make money and benefit the activist in this industry and reduces its destructive effects on the environment and the local cultures to be preserved for the future generations. Through this, sustainable tourism displays its positive role in protection of species diversity and therefore poverty is alleviated and the sustainable development will occur [2]

Types and scope of environmental effects of tourism closed depend on types, the density and the expansion of tourism. Researchers like Pigram (1980), and Travis (1982) indicated the significance of the environmental effects of tourism [3] . Naming some Iranian studies in this field, the works of Rezvani on the role of ecotourism in protection of environment [4] : Karim conducted a research on geographical and ecological effects of costal tourism and sustainable development [5] and Alizadeh examined the impact of tourists' presence on ecological resources in the villages of Torghabeh in Mashhad [6] .

2. METHODOLOGY

In the present research, first the concept of sustainable development and different dimensions of relationships between the environment and tourism as well as species diversity was examined. Second, the positive and negative effects of tourism industry on the environmental condition in the study area were evaluated and finally, existing strengths and shortcomings in the region for tourism attractions were taken into consideration. Moreover, the climate data of the region, geographical features and day and night climatetourim conditions of Semnan using the indexes like comfort index of Olegi were measured. To do this, maximum and minimum temperature, and relative humidity during a ten year period were used.

2.1. Sustainable Development and Tourism Planning

Today's, tourism has been recognized as one of the most profitable industries in the world and many countries have been struggling to exploit from tourism industry as much as they can. The economic benefits of this industry are considerable to extent that includes approximately 7% of the world capital [7]. Currently, all countries either developed or under developing or undeveloped have agreed that any development takes place only through planning. Thus, the effective factors on the development prices are divided into two categories as:

1) Natural factor: climate, geographical location, area, access to sea.

2) Human factors: population, economic system, social system, technology [8].

Generally speaking, planning can resolve conflicts may be created in the prices of development. Development of tourism in one hand can leave positive economic, social, cultural and environmental benefits in the tourist society and on the other hand may bring some negative effects in all aspects especially the ecological aspect. In fact, the tourism industry is an integration of different activities, services and industries, thus it contains some special elements are divided into two general categories:

1) Tourism resources: including natural, cultural, historical, and human resources that they individually make various forms of tourism. 2) Infrastructures: refer to all infrastructural and super structural constructs of the country and generally contain communication, health, transportation systems, hotels, restaurants, shopping centers, recreation centers etc. [9]. Figure 1 shows sustainable urbanism infrastructure and Figure 2 shows the process needed to achieve this.

2.2. Interrelationships of the Environment and Tourism

Turning the new century, fundamental structures of communities have been revolutionized and illogical exploitation of the nature has forced man to look more precisely at the disaster of excessive exploitation of the nature and has moved him to plan more comprehensively and appropriately in this regard. Then, it finds a solution for protection of the environment in one hand and a safe exploitation from the nature.

However, the first scenario in this situation is to trust the local for protection and improvement of ecotourism in any country and using their help the nature will be reserved and the existing facilities in the region will be properly identified. The second solution is to inform the local to back the nature and use it correctly. As a matter of fact, a relationship exists between tourism and the ecology in three forms as following:

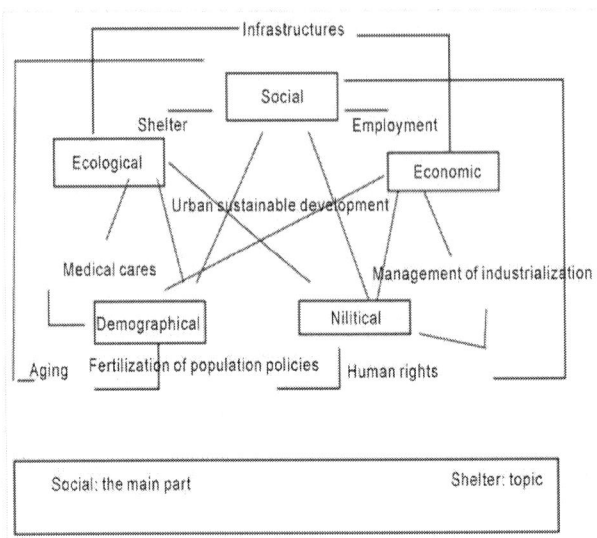

Figure 1. sustainable urbanism: main parts and the significant subjects.

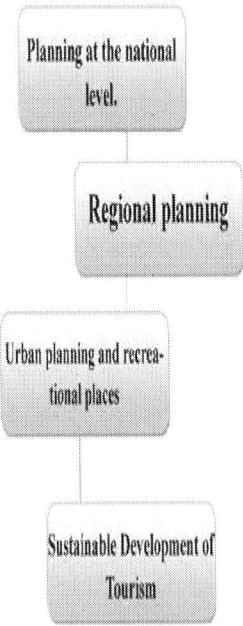

Figure 2. the process of planning for achieving to sustainable development of tourism.

1) Some forms of physical ecology are tourists' attractions.

2) Expansion of tourism and use of tourists from one specific place will leave environmental effects.

3) Facilities and infrastructures of tourism make a part of created environment.

The tourism activities also can establish plans contribute to protection of the environment. Therefore, we see that tourism is able to interact with the environment in different forms and takes outstanding steps in preserving the environment. Some of these effects are listed below.

- Tourism can make a flow of coming and going to the attractions ad does some activities for supporting ad managing these areas.

- Tourism can attract the public attention to historical and ancients attractions and helps to perform preservation and reconstruction plans for these places with association of investors.

- Tourism absorbs the public attention to the species diversity specially the endanger ones.

- thanks to significance of sun rise as one of tourist attraction and motives, this phenomenon can display the risk of resorption and loss of ozone layers and force planners and decision makers to do activities in this regard.

- The established infrastructures for the tourism industry are beneficial for the local as well. They are railroads, roads, airports etc.

- Tourism provides opportunity, and ground for preservation and maintenance of traditional transportation systems, old steam trains and ships that is a strategy for proper saving and exploitation.

Tourism can act efficiently in development, modernization, and beautification of the industrial places, mines and sea shores that have been abandoned for years after the exploitation period [10]. Therefore, due to plan for tourism considering the environment, understanding the effects of tourism on the environment will be the very first important problem [11].

2.3. Introduction of the Study Region

The city Semnan is located between three cities of Damghan, Garmsar, and Mehdishahr at longitude 53 degrees and 23 minutes and at latitude 35 degrees and 34 minutes. The average height above sea level is 1130 m. The distance to the city Tehran is 216 km, and it is connected to the transcontinental railroad Tehran-Mashhad [12]. The climate in summer is hot and dry and winter is relatively cold. Rainfall occurs in the cold season and the average annual rainfall is 140 mm. The average annual temperature is 7.17°C, while the absolute maximum temperature is 5.44°C and absolute minimum is −4.6°C. The seasonal river of Roudbar is located on northwest of the city that originates from the Alborz mountains and crosses the city Mehdishahr puts into Dasht-e Kavir [13] . Location in respect other cities is shown in Figure 3.

2.4. Tourist Attractions of the Region

As a historical city, Semnan is located on the road to the famous Silk-Road and has lots of historical and cultural attractions represent the rich culture and ancient history in the region. Of the most significant attractions are Darband cave, the historical house of Tadayon, the Rajabi house, Pahneh bath that has been used as museum currently, historical timcheh of Pahneh, the Arg gate, the great mosque, the Imam mosque. and the Imam mosque. Some historical and cultural elements of the city Semnan is referred in Table 1.

3. ANALYSIS CLIMATE TOURISM (CLIMATOURISM) OF SEMNAN

One of the most significant factors in relation to standards of a good life is taking into consideration the quality of comfort properties in the place. The human ecology depends greatly to the thermal equilibrium of man's body temperatures with the environment temperature. The ecological comfort equilibrium happens only when the balance between absorption and desorption temperature of the skin and surrounding environment is created and makes Ronnie body temperature fix at 37°C [13]. To assess these indexes, the data of years 1982-2004 were extracted from the synoptic station of the city Shown in Table2

3.1. The Gioni Diagram

In 1969 Gioni introduced the bioclimatic diagram of buildings. This diagram clearly shows the comfort zone of the man in relation to humidity and temperature. In fact, different buildings and living in the, do not provide comfort in a way that the comfort conditions inside the building is measured via a diagram named eco-building. For instance, the results of computations indicate that in March during the day time buildings are heated via sun temperature, but in at night there is no need to heating instruments [14]. Other results are shown in Figure 4.

3.2. Olegi Comfort Diagram

Using the climate comfort diagram (Olegi), it would be possible to determine duration of annual cold and heat of different cities and regions and to see how much extreme the thermal conditions of different areas are. Then through these results we can specify a mechanical system and rate of the need each residential building may have in terms of humidity, heat and cold air [15] . Other results are shown in figure 5.

Figure 3. Geographical location of Semnan city.

Table 1. historical and cultural elements of the city Semnan.

Mosques	Tourist attractions of Semnan
Government sites and castles	Mosque Zavqan-Imam (Shah) Mosque-Semnan minaret of the grand mosque
Caravanserais	Gate of Semnan citadel-building capital (Aboozar Field)-Zavqan castle-ancient fortress, fort-Fort parchment
Shrines	Yahya ibn Musa al Shrine-Shrine of Ali ibn Ja'far-Shrine of Ali ibn Ashraf-Sir Sar Tomb-Tomb of the Prophets
Tomb	Hakim Elahi-Pir Najmuddin-Darvish Mahmoud
Markets	Grand market of Semnan-Sheikh Alaa al Doleh Market
Bath	Across bath-bath Nasar-Bath Ghelli

Table2 Architecture compatible with climate condition for the gioni index.

	segment	Architecture ruling
1	1-3,8-25	Thermal exchange must be reduced through building walls
2	1-3,22-25	To prevent from air penetration from windows and inconvenient joints
3	4-25	Sun heat must be minimized as much as possible
4	1-3	The sun heat must be exploited.
5	8-13	The cross ventilation (curran) have to used for rooms.
6	18-20,15-16,13,9-10	Cooling due to evaporation of surface
7	12-19,9-10	Using cooling from long wave radiation-induced wall heat
8	21-23	Using from mechanical cooling system
9	22	Using from mechanical cooling system and dehumidifier
10	4-25	Prevent sunlight from entering the building

Table 3. Comfort and tourism conditions in Semnan.

No.	Month	Comfort conditions of tourism and ecology protection
1	April	In March there is no need to use of fossil energy and there is comfort inside and outside the buildings, therefore it is a good time for tourism in harmony with the environment.
2	May	During day there is no need for heating, the comfort is possible and use of natural ventilation is good. Somewhat appropriate for tourism, taking into account the requirements and installations is proper.
3	June	In June the need for fossil fuel and produce no pollution and natural conditions for tourism is in harmony with the environment
4	July	
5	August	In August, the energy required to cool the air at night is not on iodine physical tools used in creating comfort and energy consumption
6	September	
7	October	On the use of natural ventilation air can be brought into the comfort zone, and at night there is no need to heating appliances.
8	December	In most cases, using the sun's heat can be secured from the house, but at night it can be used to heat items. In late November, the comfort zone is located in tourism due to fossil fuel use and energy use will be in harmony with the environment
9	November	There was a sense of comfort and indoor use heat sources are necessary. Outside of the building is very cold weather and atmospheric conditions governing tourism facilities and no visit would be
10	January	There was a sense of comfort and indoor use heat sources are necessary. There are therefore fossil fuel consumption and pollution in the environment. Very cold air outside buildings and tourist facilities and no visit would be the dominant atmospheric conditions.
11	February	There was a sense of comfort and indoor use heat sources are necessary. Exploitation of fossil fuels and energy this month is high and will cause pollution of tourism. Very cold air outside buildings and tourist facilities and no visit would be the dominant atmospheric conditions
12	March	Due to the thermal comfort zone to heat environment and fossil fuel heating appliances need to be warm and outside the building envelope is required. Meanwhile, in late March of air stiffness lessens and comfort is provided.

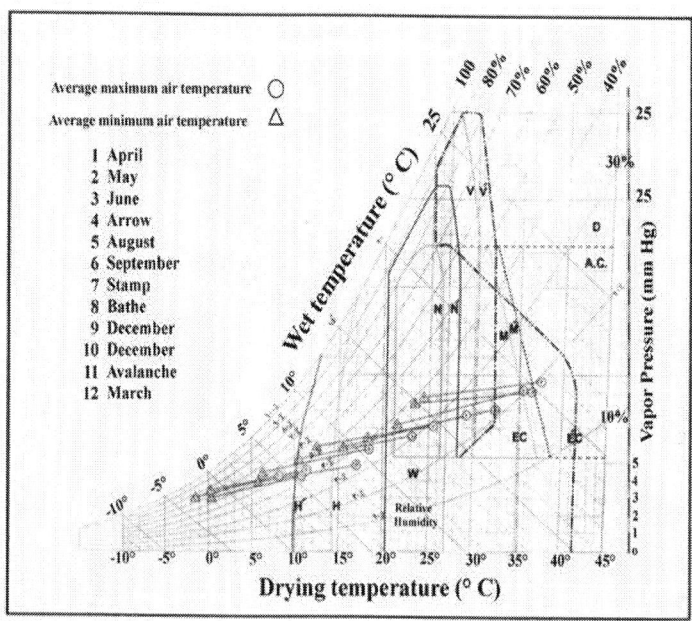

Figure 4. the bioclimatic diagram (Gioni) for Semnan city during statistical period.

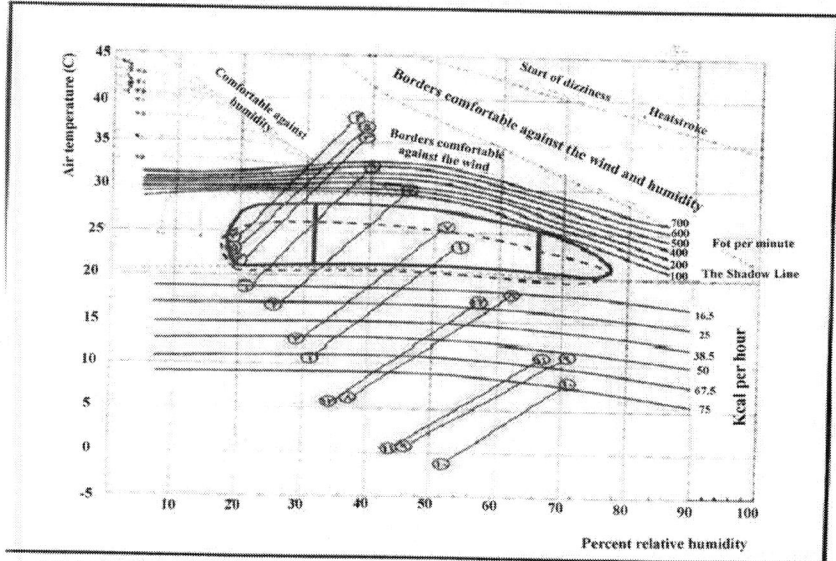

Figure 5. drawing the Olegi comfort diagram in Semnan.

4. CONCLUSION

In Semnan, proper conditions are available for use of natural energy to provide climate comfort and through proper designing of direction and intensity of sun light, type of materials, efficient use of windows etc. create appropriate use of natural energies and prevention from waste of energy and decrease of costs related to the heat and cold. The equilibrium state of ecological comfort occurs only when the balance between absorption and de Sorption. Temperature of the skin and surrounding environment is created and makes Ronnie body temperature fixed at 37°C [13]. The climate conditions of the region also affect the architecture style and the type of materials directly. All principles of sustainable development have to be embodied in a full process that leads to establishment of a healthy environment [7]. Comfort and tourism conditions in Semnan are referred in Table 3.

REFERENCES

1. Deputy of Economic Planning and Affaires, Evaluation of Tourism in Iran and Islamic Countries.
2. www.chtn.ir/Webforums/Fa/Article/ArticleInfo.apx?ID=1859
3. Edward, I. (1999) Tourism Planning: An Integrated and Sustainable Development Approach. VNR Tourism and Commercial Recreation Series, 208.
4. Rezvani A.A. (2001) The Role of Ecotourism in Protection of the Environment. Journal of Environmental Studies, 115-122.
5. Karimi, T. (2004) Geographical and Ecological Effects of Coastal Tourism and Sustainable Development. Education Office of Abhar, Abhar, 1-18.
6. Alizadeh, K. (2003) The Impacts of Tourists on the Environmental Resources, Case Study: Torghabeh District in Mashhad. Journal of Geographical Researches, 35, 55-70.
7. Goharian, M.A. and Ketabchi, M. (2005) International Tourism. Amir Kabir Publication, Tehran.
8. Zerang, A. (20060 Geographical Properties of the Developed Countries. Islamic Azad University, Tehran/
9. Shadi, M.A. (2008) Transportation the Missing Link. Journal of Terminals of the Municipalize,
10. Batra, G.S. and Chawla, A.S. (1995) Tourism Management, A Global Perspective: Department of Business Management. Deep & Deep Publication, New Delhi, 95-96.
11. Ranjbarian and Zahedi (2000) Tourism Planning in the Local and National Scale. World Tourism Organization.

12. Jankens, C. (1995) Translated by Farhad Mortezaei. Marandiz Publication, Tehran.

13. Golparvar, N. (2008) Man, Nature and Architecture. Jahan Hele, Tehran.

14. Kasmaei, M. (1984) Climate and Architecture, Center for Housing Researches, 20.

15. Kasmaei, M. (1989) Zoning and Guide for Climate Designing: Dry and Hot, Center for Housing Researches.

CHAPTER 9

Study on the Authenticity of Intangible Cultural Heritage —Take Guangdong Nanhai Boluodan Temple Fair as an Example

Jun Hu[1,2], Xiaona Feng[3], Mu Zhang[2*]

[1]Management School, Jinan University, Guangzhou, China

[2]Shenzhen Tourism School, Jinan University, Shenzhen, China

[3]Shenzhen Radio and Television University, Shenzhen, China

ABSTRACT

Authenticity is a difficult academic problem in the tourism. The authenticity of the cultural heritage represents the value and appeal of heritage. With the acceleration of intangible cultural heritage (ICH) tourism development, scholars have been calling for protecting the authenticity of ICH at the same time of the development of ICH. What is the authenticity of the content in the end and how to protect are lack of being studied. Thus, the aim of this study is to research through reviewing literatures and finding characteristics of ICH, to examine the means and differences of tourists' perception about the authenticity of ICH, so as to achieve the goal of protecting and developing ICH better. The study investigated tourists with the experience of visiting Bo Luo Dan Festival temple, and used SPSS to carry on the quota date statistical analysis. This study researched the main factors which affected the perceived of authenticity, and the relationship between the factors and satisfaction. The study got two core factors that had an impact on perceived authenticity, and told that there was a significant positive impact between the authenticity of ICH and the satisfaction of ICH tourism.

Keywords: Intangible Cultural Heritage, Authenticity, Protection, Percei

1. INTRODUCTION

It is hard to protect the incorporeity, inheritance mechanism as well as the living environment of intangible cultural heritage (hereinafter referred to as the "ICH"). Many heritages are endangered, but the government invests a lot of money, which can only benefit the minority of heritage protection and inheritance. In this case, many enterprises pay attention to the economic value and travel value of ICH, and put forward the industrialization road of ICH tourism development. The ICH tourism development aims to seek the balance point of ICH development and protection, and use it moderately and reasonably without destroying its authenticity; and the ultimate goal is to be continued and inherited. The folk custom is developed deeply in the ICH tourism development, but also often doubted by the public, such as the commercialization of ancient towns and streets, the appearance of fakelore and the change of traditional religious rites' connotation. In the process of tourism development, in order to attract tourists, ICH needs certain packaging and change, but it needs to be aware that the excessive packaging and catering will distort the nature of culture, and eventually lead to the loss of the ICH original cultural connotation, namely the so-called authenticity. The authenticity of ICH is the important basis to measure the effect of cultural heritage tourism development; therefore, taking ICH as the tourism resource and protecting its authenticity are worthy of further discussion. The ICH authenticity perception of tourists, as the main body of tourism activities, has reference value and guiding significance for ICH tourism development and protection.

This research expects to combine the ICH characteristics, puts forward the evaluation indexes that affect ICH authenticity, finds the core affecting factors through the perception difference of tourists, and finally verifies the relationship between ICH authenticity and ICH tourist satisfaction, in order to provide theoretical foundation for the ICH authenticity protection.

2. LITERATURE REVIEW

The word "Authenticity" comes from Greek, also translated as the truth, original nature, genuine instincts, reliability. Authenticity was originally used to describe the museum of art, and MacCannell (1973) [1] introduced the concept into the study on tourism motivation and experience for the first time. Pearce (1985) [2] and Waller (1998) [3] prove the tourists' pursuit of authenticity from the psychological perspective. In the field of tourism resource development, the authenticity requires the historical and cultural heritage to highlight its history and tradition, namely the authenticity of tourism object. For the amusement theme parks, they can inspire visitors' authenticity according to the psychological needs of tourists. The folk ICH, it is necessary to balance the authenticity of the subject and object, and hold the balance of tourism development. Halewood (2001) [4] , Shepherd (2002) [5] and Waitt (2000) [6] have different opinions about tourism's authenticity influence, and think that anything will lose authenticity as long as it is created for commercial purposes. But most scholars do not agree with this idea; Apostolakis (2003) [7] and

Yeoman (2007) [8] think that authenticity is a guarantee the travel suppliers should provide; Cohen (1988) [9] presents that when the traditional culture is abandoned by the modernization, the business purposes can help keep the traditional culture and customs; Van der Borg et al. (1996) [10] believe that strengthening the authenticity of the local culture is good for the reconstruction of traditional culture and national identity. Gao Fang (2008) [11] takes the "Yunnan Reflections" as an example, and proposes that under the condition of cultural industrialization, the cultural commercialization is not opposite to the authenticity; instead they can promote each other. Ma Xiaojing (2006) [12] points out that if the tourism commercialization is controlled by the local people, the national tourism and cultural tourism can be the path to the national cultural renaissance.

For cultural tourism products or destinations based on humanity, tourists' perceived value is mainly from the recognition of cultural tourism products authenticity, and tourists' authenticity perception of cultural tourism products is a dimension to assess the perceived value. McIntosh (1999) [13] thinks that in the process of appreciating the historical heritage, tourists' "authenticity" perception is influenced by the gender, income, education level, living standard, social status and early visiting experience. Liao Renjing (2009) [14] takes Nanjing Confucius Temple as an example, and studies the scenic spot "authenticity" that aims to meet the tourists' needs of historical cultural and commercial consumption from the perspective of tourists. Tian Meirong and Bao Jigang (2005) [15] analyze the tourist satisfaction from the perspective of authenticity. Feng Shuhua and Sha Run (2007) [16] establish the "authenticity-satisfactory" evaluation model from the perspective of tourism perception. Gao Yan (2007) [17] verifies that the perception difference has certain positive or negative influence on the tourist satisfaction.

Foreign research shows that the authenticity is the core value of cultural heritage. The authenticity has special meaning for the development and protection of cultural heritage. At present, scholars study the ICH authenticity mainly from the perspective of concept, and put forward protection concepts. Domestic research begins to use empirical analysis to study the perception of cultural heritage authenticity, but there is not much research result of the ICH authenticity perception, and there is even little research on the ICH authenticity protection through the perception.

3. EMPIRICAL STUDY

3.1. Research Case

Guangdong Nanhai Boluodan is the unique traditional folk festival in Guangzhou and even the Pearl River Delta Region, which is the biggest local folk temple fair, as well as the only sea sacrifice activity, containing the most representative folk cultural elements in Guangzhou. In June 2011, the State Council announced the third batch of national intangible cultural heritage list, and Boluodan was selected as the state-level folk ICH project. In the annual Boluodan period, the surrounding villagers pray for peace according to the ancient folk traditions and customs. At the same time, hundreds of thousands of villagers and devout men and women in the Pearl River Delta Region come to Nanhai Temple, bless, sightseeing and go shopping. But only building the temple fair into a tourist project but ignoring the spiritual construction deviate from the original idea of buildingGuangdong well-known cultural project. The successful ICH development can both satisfy tourists' visual and auditory needs, but it should also pay attention to the cultural atmosphere building. So Boluodan is selected as the research object of empirical study.

3.2. The Selection of Tourists' Perception Index

The concept of ICH authenticity has no unified standard. To facilitate the operation, integrate a number of research results at home and abroad, this study adopts the authenticity connotation of three heritages: 1) the objective authenticity involves the architectural style, architectural image perception, unique internal design of local and street, especially the authenticity perception of natural crafts; 2) the constructive authenticity refers to the tourists' faith expectation, preferences, image and consciousness; 3) the existence focuses on tourism subject, including the tourists' feelings and emotions, such as the unique spiritual experience and the understanding of history and culture. Based on the three aspects, analyze the influence factors of Boluodan (see Table 1).

3.3. Research Design and Field Investigation

The study design, through the questionnaire survey of tourists' authenticity perception of Boluodan Temple Fair, assesses the development effect of Boluodan Temple Fair, thus offering opinions and suggestions for the development and protection of Boluodan Temple Fair. The first part of the questionnaire understands tourists' experience, expectations and motivation of Boluodan. The second part uses Likert 5 subscales to measure the visitors' perception degree of authenticity; 1 means completely disagree, 2 means agree a little; 3 means whatever; 4 means agree much and 5 means completely agree. The third part collects tourists' gender, age, education, occupation, income and population characteristic information. The research adopts the one-to-one filling, extracts appropriate proportion according to the visitor flow during Boluodan Temple Fair, and conducts field survey. There are a total of 600 questionnaires, and 565 valid questionnaires from the effective rate of 94.2%.

4. STATISTICAL ANALYSIS

According to the survey of tourists' authenticity perception of Boluodan Temple Fair, the author adopts SPSS 19.0 as the statistical software and handles the measurement data.

Table 1. Influence factors of Boluodan.

No.	Item	No.	Item	No.	Item
1	Historical site	5	Boluozong	9	Commercialization
2	Surrounding scene	6	Special performance	10	History legend
3	Craftsmanship	7	Detailed publicity material	11	Religion and spirit
4	The lack of Boluoji's characteristics	8	Atmosphere	12	ICH title

4.1. Sample Feature Analysis

This research questionnaire collects 565 valid questionnaires. The data statistical results show the moderate sex ratio. There are more visitors aged 21 to 50 with a total of 76.3%; there are more students, administrative departments' staffs and company employees; there is more wage-earning class; there are more people with education of high school to undergraduate. Visitors of Boluodan Temple Fair are mainly residents in Guangzhou and the Pearl River Delta, as well as the strangers working there.

The blessing is the main purpose of tourists visiting Boluodan Temple Fair; experiencing folk custom and relaxing are the second purpose; buying traditional arts and crafts and gathering is also one of the purposes. From the perspective of tourism, made 80% of visitors travel together with families and friends, and individuals and those with the units account for less. First-time visitors account for 61.4% and those for more than three times account for 16.6%.

From the perspective of the tourist perception and understanding, tourists understand Boluodan Temple Fair from relatives and friends, past experiences and television publicity, followed by books and newspapers and network information. Those through the travel brochure account for less. 9.2% of visitors know much about Boluodan, 61.4% know a little and 29% don't know it.

4.2. Impact Analysis of Population Characteristic Variable on Boluodan Authenticity Perception

According to Boluodan tourists' gender, age, income, education level and occupation difference, analyze the historical site, surrounding scenery, promotional materials, Boluoji technology, new material, performance, atmosphere, commercialization, history and legends understanding, religion and spirit, ICH variables and obtain the following results by data analysis:

1) Tourists' gender has great influence on the perception of Boluodan

Through different gender tourists for Boluodan authenticity perception of the average (mean) analysis, men and women tourists for the Nanhai Temple historical authenticity (male mean = 4.02, female mean = 3.88), folk performances of traditional (male mean = 3.83, female mean = 3.79), and the overall evaluation of Boluodan (mean = 3.77, female mean = 3.72), all aspects' authenticity perception is generally higher. By independent samples T test, observe the male and female tourists' inspection index, and find that in terms of perception of the surrounding landscape, men and women have significant differences at P = 0.001 level, and no great differences in other aspects of authenticity perception (Table 2).

2)   People in different age grades has different authenticity perception for Boluodan

Because of the different degree of knowledge of history and culture and tourism purposes, 21 - 34 age tourists and 35 - 50 age tourists, in terms of authenticity perception of historical sites exist a significant difference. Authenticity perception in the surrounding scenery, 21 - 34 age groups of tourists to the scenic surroundings of the demand is higher. In terms of production technology of authenticity perception, over 65 tourists have a pretty good idea of the traditional process of Boluoji; 35 - 50, 51 - 64 age group the making craft of visitors think of the Boluoji from inheritance, the representative of the traditional culture; 21 - 34 age tourists don't know Boluoji and craftsmanship, they are not sure whether the traditional the process. In the aspect of the production process of new technology of Boluoji, 21 - 34 age tourists think the appropriate changing the material Boluoji is a sign of advancing with the times; while visitors under the age of 20 groups for students, they tend to keep the traditional production craft process. In terms of authenticity perception of Boluoji, under the age of 20, tourists don't think eating Boluozong in Boluodan is a reflection of tradition; 20 years of age or older tourists are agree that the taste of the brown has distinguishing feature very much, is the expression of the continuation of traditional customs. In terms of performance of the traditional perception, over 65 visitors to traditional folk toward the king, archaize offerings known such as the sea; tourists and 21, 34, 35 to 50 pages of traditional folk authenticity of cognition degree is not high. Religious and spiritual experiences, under the age of 20 visitors get to meet spiritual or religious, significantly higher than other age groups of tourists.

Table 2. Different gender tourists' perception of Boluodan.

Measurement project	Male	Female	T	P
Historical site	4.02	3.88	2.424	0.789
Surrounding scene	3.47	3.59	2.428	0.001
Detailed publicity material	3.62	3.60	−1.581	0.941
Craftsmanship	3.59	3.56	−1.585	0.787
Boluoji	3.40	3.35	0.278	0.292
Boluozong	3.68	3.65	0.278	0.347
Special performance	3.83	3.79	0.383	0.927
Atmosphere	3.91	3.87	0.780	0.982
Commercialization	3.76	3.78	0.326	0.405
History legend	3.45	3.48	0.326	0.572
Religion and spirit	3.25	3.30	0.631	0.166
ICH title	3.77	3.72	0.631	0.801

3)   Different educational backgrounds cause different authenticity perception for Boluodan

In combination with single factor variance, and Pearson correlation coefficient with different degree of cultural tourists to Boluodan authenticity to study the difference of perception, different education degree of tourists in the traditional production process, performance characteristic, atmosphere, and commercialization and understand the history and legend, there is no significant difference. According to the calculation of Pearson correlation coefficient, the higher the level of education, the interest and understanding of cultural things ability increases, the clearer perception of the reality of the Nanhai Temple (β = 0.101*). Education level and is negatively related to religious and spiritual experience, namely with the increase of the level of education, the tourist facilities, the environment, also meet the requirements of the tourist commodities is higher, the religious and spiritual experience to meet the lower (β = −0.141**). On the question of whether Boluodan deserves the title of provincial intangible, negative correlation with cultural level (β = −0.095*). Mainly and tourists experience and the spirit of the cultural level, the tourists' satisfaction degree, spiritual experience indirectly reflect the cultural level, the higher the perception of things may be more in-depth, highlights for high school and technical secondary school and college and university culture level of significant differences between the tourists. Revisit directly reflect the degree of satisfaction, the cultural hierarchy levels showed a significant negative correlation with revisit (β = −0.173**), cultural degree is higher, lower tourists' satisfaction of Boluodan will be.

4) Different occupations cause different authenticity perception for Boluodan

Through the analysis of variance of the variables under different career multiple inspection, found that different professional authenticity perception of tourists in the craft tradition, atmosphere, whether to revisit or recommend friends to exist significant differences on the other part does not exist significant differences. Self- employed people in the authenticity of historical sites and the surrounding landscape perception of average evaluation, significantly lower than that of other professional groups. In production process of the traditional ways, retirees, evaluation of traditional process than in several other professional groups, and retired employees in the performance of the traditional, commercial, understanding history and legends, religious and spiritual experience, the posthumous title of authenticity perception level of average is among the highest in different occupations, shows the familiar retiree of traditional culture and nostalgia. However, retiree recognition of commercialization degree is higher than other occupational groups. In the actual interview, most of the retired people think under the impact of the modern western civilization, modern commercial makes the revival of traditional culture, they said about the commercialization of this, the same Cohen (1988) [9] put forward the point of view is consistent. Because retirees higher cultural awareness and spiritual experience, their willingness to revisit and recommend friends to come to the strongest. In terms of authenticity perception of new technology of Boluoji, farmers lower than other occupational groups, this is mainly because farmers are nearby villagers, even the production process of inheritance, new materials to make Boluoji more easy to make, at the same time, in their view, the Boluoji still exists, so they do not think the new material influence the authenticity of the Boluoji. The commercialization of the Boluodan Temple Fair introduces foreign shops more, less vested interests of local farmers, so their recognition of commercialization is the lowest among all occupational groups.

5) Different income causes different authenticity perception for Boluodan

Traditional different tourists in the process of income, the authenticity of Boluozong awareness and participation in the next year or in the willingness to recommend friends to significant difference, there is no significant difference in other ways. According to Pearson correlation coefficient, the $P = 0.01$ level, different income groups of Boluozong ($\beta = 0.116^{**}$) there is a significant positive correlation, the authenticity of the perception that with the increase of income, the higher the authenticity perception of Boluozong. According to field survey, the traditional production of Boluozong is about 3 times the price of other rice dumplings, high income repre- sents the higher consumption ability, combining with the single factor analysis of variance, the outstanding performance for below 2000 income group with three other age difference of perception level, the higher the income, the more obvious differences. Under $P = 0.05$ level, different income groups in detail of publicity materials (beta $= 0.092^{*}$), production industry of traditional authenticity perception (beta $= 0.106^{*}$), there is a significant positive correlation, as incomes have increased, detailed degree of publicity materials and Boluoji processing technology of traditional authenticity perception degree is higher. High income represents high-cultural level, to some extent in the process of the visit to the temple fair of publicity materials to introduce more attention, at the same time higher education make tourists interested in the tradition of Boluoji more.

4.3. Principal Component Analysis

After factor analysis for the first time, extract three common factors. But due to the third factor contains only the new technology of making Boluoji. Item number only level 2 or 1, affect the content validity, therefore the level and the item are deleted, and authenticity perception of 11 factors for a second factor analysis.

The bigger the KMO value is, the greater the common factors between variables are, the more suitable for factor analysis. The survey data of KMO value of 0.960, suitable for factor analysis (see Table 3).

In addition, from the Bartlett ball test chi-square value of 2736.124 (DOF for 55), reached a significant level, represents the correlation matrix of the female group, there is a common factor between suitable for factor analysis. Extraction factor based on eigenvalue is equal to 1, a total of two factors, the variation explained variance of 73.694%. In order to clearly understand each factor on the 2 kinds of common factors distribution, maximum variance orthogonal rotation factor matrix. After rotating the composition of the coefficient of the size of the said factors and obtain the correlation factors, the stronger the correlation coefficient, the greater the said. The extraction of factors should be considered after rotation factor load values at least those variables should be greater than 0.50.

It can be seen from Table 4 that: extract 2 factors from 11 key elements, 2 factors overall explains 73.694% of variation in the data. Two factor analysis at the heart of the tourist authenticity perception influence factor, the factor 1 includes: religious and spiritual, understanding of history, legends, the posthumous title, atmosphere, historical sites attract, surrounding scenery and commercialization; factor 2 include: craft tradition, Boluozong, publicity materials and performance characteristic in detail and so on four elements.

Table 3. KMO and bartlett's test.

Kaiser-Meyer-Olkin sample degree		0.960
Bartlett's test	Chi square distribution	2736.124
	DOF	55
	Significance	0.000

Table 4. Factor loading table after spindle.

Factor	Measurement project	Component	
		1	2
F1	Religion and spirit	0.870	
	History legend	0.860	
	ICH title	0.794	
	Atmosphere	0.789	
	Historical site	0.707	
	Surrounding scene	0.618	
	Commercialization	0.600	
F2	Craftsmanship		0.819
	Boluozong		0.815
	Detailed publicity material		0.591
	Special performance		0.584

Note: the table only contains the load above 0.5.

4.4. Linear-Regression Analysis Based on the Tourists' Authenticity Perception Satisfaction

Establish two new variables F1 and F2 by the method of principal component analysis to reflect the tourists to the authenticity of sensory information. On this basis, in order to "next year will take part in or recommend friends to" as indicators of satisfaction, F1 and F2 as independent variables. Multiple linear regression was carried out on the degree of satisfaction. Establish tourist authenticity perception functional relations between the influencing factors and the degree of satisfaction, and verify the significance of each variable. By the fit of the model test results (Table 5), the regression model of Sig. The value is 0; shows that the model is significantly related. The adjusted R2 is 0.855, has high fitting degree. Fitting is presented in Table 6 are not standardized and the standardized regression coefficients (including the constant term), and through the T test method for fitting of test results, the Sig. The value is 0, has significant statistical significance, can build a linear model.

According to the non-standardized coefficient in Table 7, the regression results are available, and the linear relation of F1 and F2 and satisfaction is as follows:

$$Y = 1.144X_1 + 0.622X_2 + 3.237$$

Therefore, two common factors F1 and F2 have a positive influence on tourist satisfaction; the F1 class public factor's influence on the degree of satisfaction is 1.144, while that of F2 positive impact is 0.6222.

5. CONCLUSIONS

Authenticity of ICH has an important influence on the development of ICH as the tourism resource, and the authenticity the value and appeal of the heritage. To protect the heritage authenticity is a problem that is worthy of discussion, as well as one of the important ways to protect ICH. The research is based on the cognitive theory, combines the characteristics of ICH, and studies the protection of ICH by studying the approach and difference of tourists' cultural heritage authenticity perception.

The research takes the state-level ICH Boluodan Temple in Guangdong Province as an example, through the field survey of the Nanhai Boluodan Temple folk festival, uses SPSS19.0 to analyze the data, and finds that tourists' different gender, age, educational level, occupation and income have different influences on Boluodan authenticity perception. Through the principal component analysis, it can be seen that two core factors have positive influences on tourist authenticity perception; factor 1 includes: religion and spirit, the understanding of history, legends, ICH title, atmosphere, historical sites, surrounding scenery and commercialization; factor 2 includes: craft tradition, Boluozong, publicity materials and performance, etc.

Table 5. Assessment of model fit.

Model	R	R square	Adjusted R square	Std. error of the estimate	Change statistics				
					R square change	F change	df1	df2	Sig. F change
1	0.925	0.856	0.855	0.536	0.856	883.751	2	297	0.000

Table 6. Analysis of variance table.

	Model	Sum of squares	df	Mean square	F	Sig.
	Regression	507.003	2	253.502	883.751	0.000
1	Residual	85.194	297	0.287		
	Total	592.197	299			

Table 7. Regression analysis results.

	Model	Unstandardized coefficients		Standardized coefficients	t	Sig.	95.0% confidence interval for B	
		B	Std. error	Beta			Lower bound	Upper bound
	(Constant)	3.237	0.031		104.673	0.000	3.176	3.298
1	F1	1.144	0.031	0.813	36.944	0.000	1.083	1.205
	F2	0.622	0.031	0.442	20.067	0.000	0.561	0.682

Therefore, it can be seen that the tourists' perception of ICH authenticity not only contains the perception of heritage physical carrier, but also includes the perception of traditional culture heritage; in addition, the tourists' perception mainly comes from their own experience and knowledge, as well as the planning and arrangement of tourism project. So taking Nanhai Boluo Temple as a tourism resource can be designed and treated according to the general rule of tourists' perception, and its basis lies in that tourists' expectations, evaluation and attitude towards ICH will affect the tourist experience, affecting the tourists' satisfaction for folk ICH.

The study also needs more of ICH as examples to verify, and this research hopes that this conclusion can provide a theoretical basis for the authenticity protection of ICH.

ACKNOWLEDGEMENTS

This paper is supported by the Higher Education and Teaching Reform Project of Guangdong Province in 2012 with Grant Number 2012069; Project of Scientific Planning of Guangdong Province with Grant Number 2012B031400008, and the key project of 13th teaching reform research of Jinan University.

REFERENCES

1. MacCannell, D. (1973) Staged Authenticity: Arrangements of Social Space in Tourist Settings. American Journal of Sociology, 79, 589-603.

2. Pearce, P.L. and Moscardo, G.M. (1985) The Relationships between Travelers Career Levels and Concept of Authenticity. Australian Journal of Psychology, 37, 157-174.

3. Waller, J. and Lea, S.E.G. (1998) Seeking the Real Spain—Authenticity in Motivation. Annals of Tourism Research, 25, 110-128.

4. Halewood, C. and Hannam, K. (2001) Viking Heritage Tourism: Authenticity and Commodification. Annals of Tourism Research, 28, 565-580.

5. Shepherd, R. (2002) Commodification, Culture and Tourism. Tourist Studies, 2, 183-201.

6. Waitt, G. (2000) Consuming Heritage-Perceived Historical Authenticity. Annals of Tourism Research, 27, 835-862.

7. Apostolakis, A. (2003) The Convergence Process in Heritage Tourism. Annals of Tourism Research, 30, 795-812.

8. Yeoman, I.S., Brass, D. and Mcmahon Beattie, U. (2007) Current Issue in Tourism: The Authentic Tourist. Tourism Management, 28, 1128-1138.

9. Cohen, E. (1988) Authenticity and Commoditization in Tourism. Annals of Tourism Research, 15, 371-386.

10. Van der Borg, J., Costa, P. and Gotti, G. (1996) Tourism in European Heritage Cities. Annals of Tourism Research, 23, 306-321.

11. Gao, F. (2008) Cultural Commercialization and Authenticity Relationship in the Ethnic Tourism Development—Take "Yunnan Reflections" as an Example. Journal of Baoshan Normal College, 3, 53-56.

12. Ma, X.J. (2006) Research on Foreign Cultural Heritage Authenticity. Guangxi Ethnic Study, 3, 185-191.

13. McIntosh, A.J. and Prentice, R.C. (1999) Affirming Authenticity Consuming Cultural Heritage. Annals of Tourism Research, 26, 589-612.

14. Liao, R.J., Li, Q., Zhang, J., Lu, S.J. and Qi, Q.Y. (2009) Recreation on Tourists' Perception for Urban Historical Blocks Authenticity—Take Nanjing Confucius Temple as an Example. Tourism Tribune, 1, 55-60.

15. Tian, M.R. and Bao, J.G. (2005) Research on Visitors' Authenticity Evaluation for Musical Tourism Product—Take Xishuangbanna Dai Song and Dance as an Example. Journal of Guilin Tourism Advanced Science School, 1, 12-19.

16. Feng, S.H. and Sha, R. (2007) Primary Exploration of Assessment Model of "Authenticity-Satisfaction" for Ancient- Village Tourism. Human Geography, 6, 85-89.

17. Gao, Y. and Ling, C.R. (2007) Tourists' Authenticity Perception Differences and Satisfaction of Black-Clothes Zhuang Culture. Tourism Tribune, 11, 78-84.

CHAPTER 10

Towards Sustainable Tourism Development in Zambia: Advancing Tourism Planning and Natural Resource Management in Livingstone (Mosi-oa-Tunya) Area

Binyi Liu, Floyd M. Mwanza

Department of Landscape Studies, Tongji University, Shanghai, China.

ABSTRACT

Over the last few decades, development policy has been dominated by mainstream economic theories that focus on economic growth to achieve sustainable development. The pace and scale of tourism growth in Livingstone (Mosi-oa-Tunya) area in Zambia have seen over reliance on natural resource utilisation by mass tourism developments. Compounded by insufficient planning and limited co-ordination and collaboration among the institutions involved in the tourism sector, tourism can have a negative impact and can create conflicts. Tourism growth in Livingstone (Mosi-oa-Tunya) has predominantly focused on the economic incentives in tourism and ignored the social perspective and impact on the local population. In general, the government agency administration structures affect the successful implementation of tourism policy and planning for sustainable tourism development. Given the fact that the limited government support, funds and appropriate knowledge in tourism limit Livingstone (Mosi-oa-Tunya) to develop as a sustainable "green" destination and remain an enormously difficult task to achieve.

Keywords: Natural Resources; Planning; Sustainable; Tourism Policy

1. INTRODUCTION

In an effort to reduce the negative impacts of conventional tourism, more environmentally and socially conscientious approaches were promoted to tourism. Typically called "ecotourism and sustainable tourism" though other terms such as responsible tourism, nature based tourism, green tourism, and alternative tourism are also used [1]. Any tourism destination without an adequate plan for development that addresses the economic as well as social and environmental functions of the industry is under prepared for the impacts of visitors, catastrophic events, and enforcing market forces [2]. Tourism requires a great deal of infrastructure including hotel road parking lots and restaurants which typically brings a number of negative consequences [1], such as increased pollution levels, the destruction of natural habitats, the displacement of natural wildlife and undesirable influences to once remote cultures [3]. As an alternative to conventional tourism, sustainability and ecotourism has continued to gain momentum over the last two decades [4]. Planning for sustainable tourism development refers to environmental preservation planning and as such includes a variety of inquiry activities and analysis to the decision for determining the direction of the development [1,5]. Tourism planning is advanced to prevent the intensive utilisation of resources in some specific areas without previous care for the preservation of the resources [6]. There has been an increasing need for landscape planners to consider methodological approaches to tourism planning and a number of techniques, principles, and examples that have evolved and recommended [7-9]. Nevertheless, the multiplicity and heterogeneity of tourism stakeholders render the process complicated [10]. A key component to the success of sustainable "green" tourism is local control in the planning, development and management of these tourism sites [11]. Arguably, Livingstone has the highest concentration of tourism activities in Zambia [12], although many authors have examined various aspects of general economic planning viability. A literature review of tourism sector shows that, with few exceptions, most studies in the past have focused on research in tourism marketing [13], and that many now incorporate new theories and concepts for better provision of tourism [14]. Bearing in mind the above, this study aims to contribute to better understanding of how tourism development planning is carried out and the inherent difficulties are associated with the implementation. It aims to explore and examine what is happening in Zambia. Therefore, this study attempts to record and analyze the factors affecting the development of the Livingstone (Mosi-oaTunya) tourism region. It compares planning and implementation of the national and regional tourism policies and strategies and attempts to show the conditions, which affect the course of implementation and cause its divergence from sustainable national policies and regional planning objectives. It considers the processes of translating objectives to outcomes and investigates why those processes fail to translate many objectives to practice.

2. LITERATURE REVIEW/BACKGROUND INFORMATION

First, Zambia's tourism industry relies on two primary assets: the Livingstone (Mosi-oa-Tunya) area and the country's wildlife estate in 19 national parks, Game Management Areas (GMAs) and game ranches Zambia Development Agency [15]. The Livingstone (Mosi-oaTunya) Falls site appeals to a large tourists range of approximately 138,830 visitors more than the safari product of 61,000 visitors in 2009 Ministry of Tourism Environment & Natural Resources & Zambia Tourism Board[16]. Livingstone tourism activities are relatively far developed, compared to other regions in Zambia, [15]. Livingstone Mosi-oa-Tunya has a large proportion of Zambia's adventure tourism capacity [17]. A recent [18] publication elaborates that between 2010 and 2030 arrivals to emerging economies will increase at double the rate. As a result, the market share of emerging economies such as Zambia has increased from 30% in 1980 to 47% in 2011 and expected to reach 57% by 2030 to over one billion international tourist arrivals [18]. However, to meet sustainable tourism, scholars argue that sustainability has largely been used conceptually as a "good idea" but has been difficult to enable through specific initiatives [4]. The task is more difficult in view of the multiple crises faced by the world. Recession, climate change, fuel crisis, food crisis, and water crisis, planning and governance become topical issues [19]. Consequently, the impetus for many of the current initiatives in tourism and international development stems from Agenda 21, a comprehensive program of action for attaining "sustainable development" in the 21st century [20].

International tourism destinations particularly those rich in biodiversity have in recent years caught attention of the global environmental movement because of resource degradation [21].

Arguably, community based planning approaches are promoted for tourism development as a prerequisite to sustainability [5,20]. Reference [22] observed that many destinations are now pursuing strategies that aim to ensure a sensitive approach when dealing with tourism. [8] refers to sustainable tourism development is an enormously difficult task to achieve in developing countries, without the collaboration of the international agencies such as the World Bank and the International Monetary Fund. Reference [23] observed the rise of sustainable tourism discourse in Southern Africa and has seen the development of a multiplicity of tourism projects packaged under ecotourism as a more sustainable form of tourism than mass tourism. The term "sustainable tourism" can mean different things to different people, often according to the position of the individual stakeholder. It is important to elaborate tourism planning with a definition of some principles of sustainable tourism [24,25]. More recently, the WTO defined "sustainable tourism" as follows: "Sustainable tourism development meets the wishes of present tourists and host regions while protecting and enhancing opportunities for the future." As [7] stated, achieving sustainable tourism is a continuous process and it requires steady monitoring of impacts, introducing the necessary preventive and remedial measures whenever necessary [25].

Furthermore, sustainable tourism is now an approach at the international level that is advocated to be adapted to prepare all types of tourism to be environmentally, socially and economically beneficial [26-28].

Tourism in Zambia

Zambia's protected area network covers 30% of the country (224,075 km^2) for which Zambia Wildlife Authority (ZAWA) is responsible [12].

The 19 National Parks covering 6587 km^2 (28%) and 32 Game Management Areas (GMAs) 160,488 km^2 in extent or 72% of the country's PA network a huge resource forms wilderness tourism supply side [29-31]. Many of the tourism activity centres on the 19 National Parks covering 63,587 km^2 (28%) and 32 Game Management Areas (GMAs) of 160,488 km^2 in extent, or 72% of the country's protected area (PA) network [29]. Many national parks landscapes and fauna form the basis for lucrative tourism and hunting industries in Zambia [30].

According to the [32] tourism activity created 44% of employment in the hotel and restaurant industry compared to 7% in the mining industry, 99% in agriculture and 66% in manufacturing. In 2010, the number of arrivals in Zambia was 815,000 and increased to 920,299 in 2011 as shown in Figure 1.

Zambia's tourism industry established itself in the 1950s, [17]. As shown in Table 1 Zambia's tourism indicators in years. There have been some significant changes in strategic and policy levels in Zambia, all of which have the potential to influence the sustainable tourism planning agenda[33].

However, the extent to which these changes have infiltrated into implementation of local government is an area that requires further investigation [34,35] and the holistic involvement of communities in effective utilisation of their environmental assets and cultural heritage. The basis for this stance stems from the factor that tourism is increasingly Livingstone major economic activity. According [16] Zambia's stake in the industry has been insignificant, but the past five years or so have witnessed a steady growth in the tourism sector, projected to deliver over 1.4 million tourist arrivals by 2015. [17]. Following Figure 2 provides more detail on the purpose of holiday visits, suggesting that more than half Zambia's holiday makers (54%) arrive with the intention of visiting Livingstone's Mosi-oa-Tunya only [36].

Zambia's major tourism supply side clusters have developed in only a few key urban and national park locations, with a strong bias to the Livingstone region that offers nearly 40% of all nature tourism [15]. The effects of uncontrolled tourism development degrade ecosystems can be negative [37,14]. Nowhere in Zambia is this more evident than in Livingstone Victoria Falls (Mosi-oatunya) tourism site [38]. This underscores the need to entrench sustainable tourism planning principles in tourism management plans well before development begins and irreparable damage are incurred [39,40].

3. RESEARCH SITE PROFILE AND CHARACTERISTICS

3.1. Study Site Location

Livingstone (Mosi-oa-Tunya) was purposefully selected for the study because; it is among the sites being developed under the Livingstone Tourism Plan in Zambia [41]. Furthermore, the area is the earliest to be established in the region as a sustainable destination [42] and as a result, was due for evaluation.

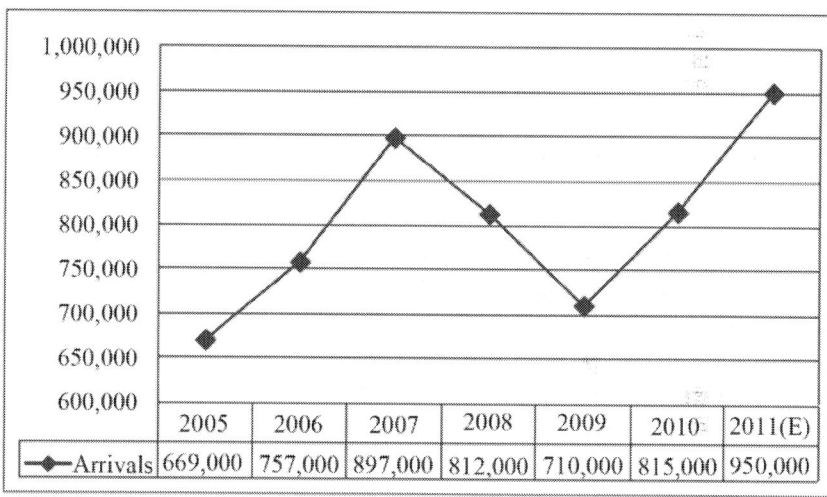

	2005	2006	2007	2008	2009	2010	2011(E)
Arrivals	669,000	757,000	897,000	812,000	710,000	815,000	950,000

Figure 1. Tourists arrivals in Zambia as of source; [16].

Table 1. Zambia's Tourism indicators in years. Source: [34].

Year	Budget Allocation in (USD)	Budget Release in (USD)	Employment Levels	GDP (%)	Tourism Earnings (USD)	Tourist Arrivals
2006	8,140,000	3,202,000	21,204	2.40%	177,000,000	757,000
2007	34,740,000	21,740,000	22,204	2.00%	188,000,000	897,000
2008	24,264,000	13,616,000	22,756	2.40%	200,000,000	812,000
2009	15,520,000		24,308	2.70%	212,000,000	710,000
2010	43,990,000	13,260,000	25,960	2.30%	224,000,000	815,000
2011	7,120,000	6,320,000	31,900		216,000,000	950,000
2012	10,520,000					
2013	12,760,000					

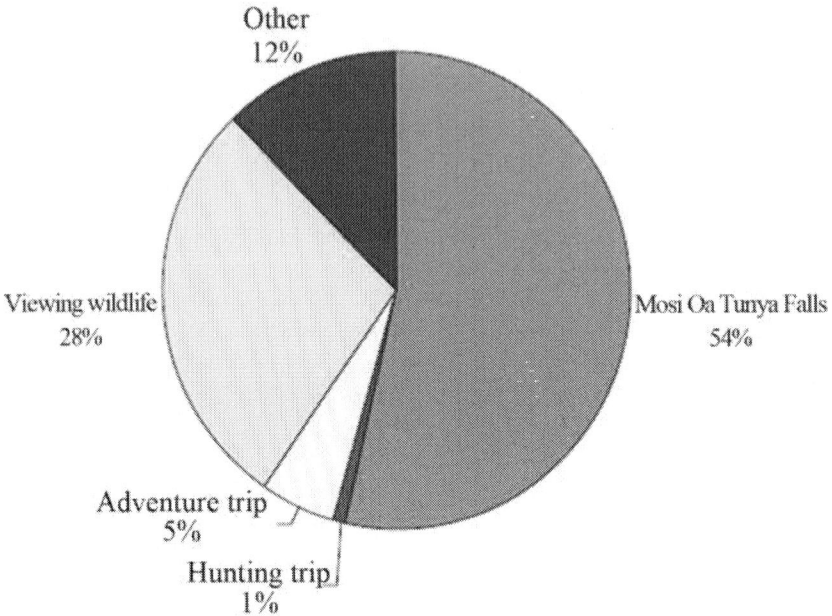

Figure 2. Tourists' destination choices in Zambia [17].

The region derives its name from the waterfall locally known as the "Mosi-Oa-Tunya" meaning "The Smoke that Thunders". David Livingstone from which the Livingstone town is named after then named the waterfall "Victoria" after Victoria, the British Queen [29]. The size of the park is about 6555 hectares or 66 square kilometers lying between latitude 17°49' South to 17°54' South and longitude 25°41' East to 25°55' East and an attitudinal range of between 900m to 925m above sea level [41].

3.2. Demography and Socio-Economic Characteristics

Livingstone Victoria Falls (Mosi-oa-Tunya) is in Southern Province of Zambia. Established in 1971 and declared a National Park in 1972 [42]. With an estimated population of 136,897 [32] and approximately has over 6000 visitors per day [16]. They contain biodiversity of global significance and is listed as Critical Sites (including Critical Habitats) [43]. There are around 4590 plants confined to this area, together with 35 endemic mammals, 51 endemic birds, 52 endemic reptiles, 25 endemic amphibians and an unknown number of endemic invertebrates [43,44]. It is approximately 20 km long and a maximum of 5 km wide [41]. It is constricted centrally to approximately 0.5 km (500m of land) [41]. Figure 3 shows the location map for Livingstone Mosi-oa-Tunya area; One of the overriding concerns about tourism in Zambia is that the tourism product relies heavily on the natural and physical environment [42,45,46]. Reference [8] highlighted that unsustainable tourism activities can affect the future viability the tourism sector, conserving of natural resources has become important through planning [47,48]. "Governments have become extremely canny in reproducing the sustainable development rhetoric without actually effecting fundamental policy and legislation changes [20]." In Zambia the need to understand the impacts of tourism has become important within a planning context because of the many statutory requirements such as the [41,49,50] and global demand for sustainable tourism [25].

The Zambian tourism sector is guided directly and indirectly by 11 pieces of legislation [51]. These tourism plans have focused merely on maximising foreign tourists' receipts and thus increasing the supply capacity of the tourism industry [14,52]. The main shortcomings are due to sectorial planning done in isolation, communication and co-operation among related bodies are sparingly weak and in most cases do not exist [34,53-56]. It is only right that development and land management is supported by a holistic planning [57-60]. It is reported that these common shortcomings in present tourism development approach pose challenges to sustainable tourism development in Zambia [35,54,61,62]. As observed by [8,63-65] the study is premised on the assumption that, local government agencies and communities to influence the sustainable tourism planning agenda. However, the extent to which tourism policies has infiltrated into local tourism agencies and communities is an area that requires further investigation. The following section outlines the methodology used to survey local agencies and communities to ascertain responses to tourism policy and planning implementation in Livingstone (Mosi-oa-tunya) area.

4. METHODS AND PROCEDURE

4.1. Research Design

The research conceptual framework was developed based research methods for assessing local authorities participation in tourism policy and planning similar to studies done by [10,63,64]. Given that the purpose of this paper is to identify and evaluate tourism policy and planning implementation for sustainable tourism development by government tourism agencies in Livingstone (Mosi-oatunya) area. A qualitative descriptive approach was employed and quantitative data where appropriate. The study used the non-probability sampling design to collect data from local tourism authorities and agencies. As cited in [66], the purposive sampling technique was found to be adequate and appropriate for such a survey research. In view of the facts given above, the purposive sampling method was adopted. Interview guides and questionnaires were the instruments used for data collection. The interviews and questionnaire administration was made to government tourism agencies (MTENR/ZAWA/ZEMA) and local Community-Based Natural Resource Management (CBNRM) representatives. In addition, institutions related to the Zambian tourism industry were also contacted for requisite information and data. To capture a significant number of tourism planners in the population sample, data were collected from tourism agencies at the Lusaka Ministry of Tourism Environment Natural and Resources, Zambia Wildlife Authority in Lusaka's Chilanga headquarters, Zambia Tourism Board Lusaka and Regional Local ZAWA branch in Livingstone (Mosi-oaTunya) area. Livingstone greater area community leaders of Community Resources Boards (CBR) agencies and popular lodges, tourism enterprises and guesthouses make the local community group. The research was conducted from 22nd November 2012 to 5th April 2013.

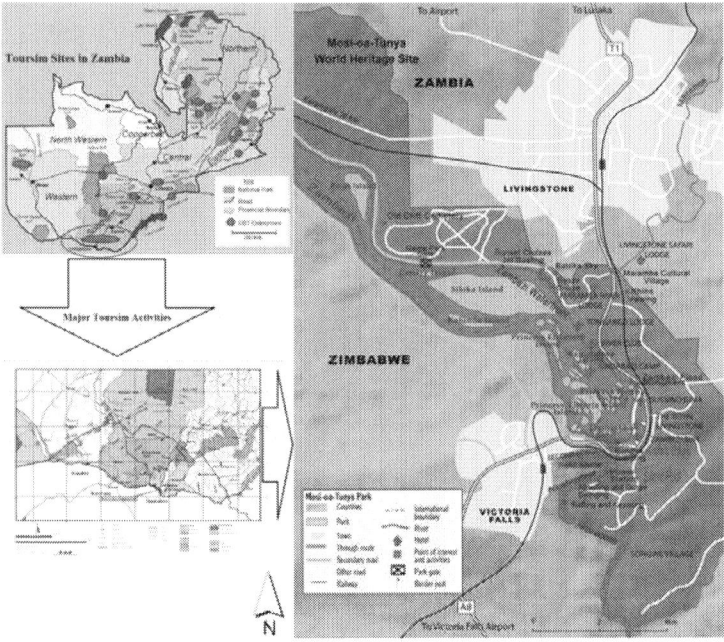

Figure 3. Map of Livingstone [35] (Mosi-oa-Tunya) of Zambia, source: modified by author (2013).

All the in-depth interviews were conducted at places of choice by the interviewees in the various departments and communities. The interviews were conducted by the corresponding author.

4.2. Data and Sources

In view of the facts given above, the purposive sampling technique was found to be adequate and appropriate because there was no sample frame of all respondents. Since it was an exploratory study, the rationale of the data collection was to ascertain government agencies and community heads' roles in tourism policy and planning participation in the sustainability framework implementation for the case of Zambia's Livingstone greater area. The survey employed three major methods, personal questionnaires, interviews of identified key actors in the Zambia tourism planning process and the local tourism businesses and local tourism authorities. The purpose however, was to generalize from a sample to the population in order that inferences could be made about the involvement of government agencies and th communities in tourism policy and planning development [67]. Secondary sources from literature review of books, journals and grey reports. Grey reports were used due to limited research publications concerned with tourism planning in developing countries such as Zambia. Past reports and any other material related to plans and policies for the tourism development of Livingstone were reviewed.

4.3. Sampling Procedure and Study Instruments

Based on the actors' group, a list of government tourism heads and local community representatives was compiled and used as a sampling frame for the selection of the respondents. A self-completion questionnaire survey was emailed to solicited sample list of local tourism authorities and planners in Zambia, who have direct and indirect involved in the Livingstone tourism master plan. A realtime online-based questionnaire website [46] was used to improve response and analysis of findings. The onlinebased survey enabled a degree of tracking and gauge findings and easy clarifications and adjustments in cases where the questionnaire was not well-defined. To encourage survey completion and to confine the aims of the survey to specific tourism planning objectives (such as identification of sustainable tourism planning implementation and development in tourism strategies), without eroding the aims of the investigation, the survey design incorporated a combination of closed and open questions which was also hosted online to improve survey responses and participation. Closed questions were utilised to gauge responses to straightforward questions, where a simple tick box suffices to assist in classification of respondents. However, recognizing the small sample specific population involved in this survey, a range of open questions were included to generate a source of more qualitative, explanatory information that can add a richer dimension to understanding responses. Hence, the fieldwork aimed to interview representatives of the major groups. It was designed using a series of semi-structured interviews with key actors. Reference [68] explained the criterion used to determine sample size is an important issue in research. The study uses descriptive data analysis and explanation, and the use of appropriate theory to help explain events [69,70].

4.4. Assumptions and Limitations

The study was based on the assumption that surveying experts with knowledge of the case study sites would be a reasonable way to obtain an up to date overview of site. However, this approach may have a few limitations, including:

- Bias of the expert's personal opinion;
- Interpretation of other stakeholder's views by a third party;
- Incomplete knowledge of the site (including of impacts, challenges and dynamics between stakeholders);
- Some experts had not been back to the site in the last two years.

This assumption and limitation were in part addressed by the desktop review of documents, which often served as a complement to the information provided by the experts.

4.5. Target Population and Sample Size

The target population for the study was government tourism agencies and local community heads or their representatives in the selected communities. A total of 85 questionnaires from 165 sent questionnaires were filled in for this particular study, 9 in-depth face to face interviews were carried out with persons involved in the policy making process and the implementation of tourism related plans. As shown in Table 3 the 85 respondent population comprised 12 local communities (CBNR) and consultants and 73 tourism planners/local authorities as shown in Table2 Surveys were mailed directly to the planning officers or agencies (MTENR, ZAWA, ZTB, ZEMA and National Heritage Conservation Commission (NHCC) who oversee on tourism planning processes and understand how tourism fits into local development plans for completion. 43 respondents completed and usable questionnaires were returned, yielding a response rate of 51%. The aim of the survey was not to produce large amounts of statistical analysis, rather to generate a picture of current levels of sustainable tourism planning at the two levels (local and regional), which is descriptive in exploring the small population size. Given that this figure represents half of all local authorities, the information the survey yielded is considered valid in providing a general picture of sector responses to tourism planning in Zambia.

4.6. Attitudes towards Tourism Development

The patterns of response provide a useful geographic spread of data, and represent a good mix of respondents. The low response rate at community level at 25%, explained primarily by the apparently delegated role of tourism planning by the local level of regional agencies and cross cutting issues, and whose main concerns relating to tourism are integrating resource management issues under (ZAWA/ZEMA/MTENR), particularly water and waste management. The respondents and their abilities to conceive the questions and to answer precisely varied not only from one group to another, but sometimes from one respondent to another in the same group. The gap in the level of knowledge, experience, and tourism planning development backgrounds and quality of information influenced the answers and the views of the respondents and thus the results obtained. At regional or both provincial and city levels, the response rates were over one half of the population (55%). Methodologically, this study suffered from the same problem as most online and mailed surveys, and while the overall response rate is satisfactory (often online and mailed 30% response rate is deemed reasonable for such surveys, for an online/ mailed survey, conventionally, a response rate of 20% is considered as a good response rate, while a 30% response rate is considered to be good [71]. It is difficult to assure the representativeness of the responses achieved. The non-respondents included 7 (Consultants & Planners) (out of 15) and 26 (local community & district planners; while for (ZAWA main branch) and regional planning offices the non-responsive figure was 9 (out of 12). Some 28 responses were received from MTENR Lusaka 49% response), while 15 responses were received from the Livingstone (54% response). Overall, the responses received provide a satisfactory sample in relation to tourism areas, population size and geography, all of which will be further elaborated in the findings. Longitudinal comparisons are only possible at the general level, given that although the same population was sampled, not all respondents answered the surveys. Lastly, it should be noted that, the names of specific government tourism departments are not given in the discussion of findings from the survey to respect the confidentiality of the research process, which was assured in the research process in order to get accurate responses.

5. RESULTS AND DISCUSSION

The findings of the survey are reported using a combination of descriptive and quantitative data given the small population, with verbatim responses to open questions to enrich the data and provide further insights. As a first step, it is valuable to recognise the scale and type of communities, areas and tourism profiles represented in these findings, particularly as such variables are useful in cross tabulating findings. The resident population of the survey areas at present population was estimated at 136,897 inhabitants at the 2010 census [32]. With reference the surrounding areas that make the largest populations, are made up of Mukuni's village on the eastern and South-Eastern border, Sekute Chiefdom (Simonga area) on the West, Imusho village to the western boundary of the park and Chief Musokotwane on the North-Western boundary are in Southern Province of Zambia, with an average population density of below 15 people km^2 residents with a total estimated population of 778,740 persons in Southern province alone [32,30].

5.1. Tourism Policies

Local and regional authorities were asked if they had knowledge of the Zambia Tourism Policy. There is no statutory requirement for a Tourism Policy, the publication of one indicates a strong community interest and local government commitment to tourism, the survey revealed that 26 tourism institutions under at the three levels of planning and implementation level have knowledge of the tourism policies and other strategies. Table 3 shows the comparison of the survey groups indicating the different types of organisational levels and knowledge on tourism policy and related strategies developed towards the tourism sector in Zambia.

The survey respondent's percentage outcome based on proximity to the study site revealed a lower understanding of the Tourism Policy. The respondents' percentage figure trends shows 33% of CBNRM respondents had knowledge of the policy and the trend rise in the knowledge of Tourism Policy and strategies by a significant rise at the main government ministry of tourism and government department agency ZAWA. This would appear to indicate that the effect of the national tourism strategy better understood at the core ministry and department and less appreciated or limited knowledge at local community level to develop and adopt strategies. Respondents with no knowledge of the Tourism Policy stated that all tourism matters delegated to the MTENR or ZAWA. Findings suggest that despite major tourism activities taking place in Livingstone and its surrounding areas, the local population have never come across the earmarked Livingstone greater area plan for sustainable development, but could be encouraged if they had one. This could explain the reason for low response from at local community level, and indicates a lack of interest in tourism development issues at this level, where tourism planning and policy issues were delegated to other bodies at ZAWA and MTENR.

5.2. The Influence of the MTENR/ZAWA Tourism Policy for Zambia (TPZ)

The majority of local tourism authority planning officers at local and government department (MTENR/ZAWA) level who had knowledge of the already existing tourism plans and had seen the Livingstone greater area plan, 83% of respondents had indicated how the Tourism Policy for Zambia (TPZ) would inform their own policy development. Responses from the local tourism authorities thought that there were emerging tourism issues that needed to be included in a future revised policy such as ecotourism certification, "green tourism", and sustainable tourism practices. Five respondents indicated the need to incorporate elements of the five year national development plans such as the [53,72,73], where appropriate to reflect particular locality and for easy implementation. While a further three stated that, they would take the plan lacked in implementation process due to institutional limitation and resources considered.

Table 2. Composition of sample population responses of key tourism actors in Zambia.

No	Type of Organisation (s);	Target Pop:165	Respondents	Response	Non-Response	% Response
	Year of Survey, 2012		**Categories of Respondents**			
1	Line Ministries (Ministry of Tourism, Environment and Natural Resources (MTENR)			40	33	55
2	Zambia National Tourist Board (ZTB)					
3	Zambia Wildlife Authority (ZAWA)	Government Officials				
4	National Heritage Conservation Commission (NHCC)		73			
5	Zambian Environmental Management Agency (ZEMA)					
6	Donor Agency	Consultants				
7	Businesses &Parastals					
8	Tourism Experts					
9	Community GMA Game Management Area Community-Based Natural Resource Management (CBNRM)	Interest and Local Groups	12	3	9	25
	Total		85	43	42	80

Table 3. Tourism government agencies & actors' respondent's percentage response, by author (2013).

Year of Survey	Percentage tourism policy @ Community (CBNRM)	Percentage Regional offices tourism policy@ (NHCC/ZAWA)	Percentage tourism policy @ (ZAWA/MTENR)
2012	33%	66%	75%

Two respondents stated that the TPZ was approved in 1997 and published in 1999, does not reflect the current organisational structures and governmental existing plans, while a further three stated that the turnaround strategy plan directly aligns with the national sustainable strategies[12]. Others commented on specific elements of the national plan and appreciated the opportunity to determine the national context and direction of tourism strategy in Livingstone Zambia and the replicability of the pilot plans such as the Livingstone greater area development plan and for a common approach to core issues as set WTO's universal tourism standards. Overall, though, the ways in which the TPZ has already influenced, or will influence, policy at a local level appear vaguely stated in many cases.

5.3. Planning for Tourism Impacts in Livingstone

Some 57% of respondents raised specific tourism issues that need redress in the next review of the Tourism Policy Zambia. The responses as illustrated in Table 4 and in some cases, respondents gave more than one reaction. The range of emerging tourism related issues raised indicate two approaches to tourism development. These approaches are not polar opposites, but do represent different perspectives on tourism activity. On one side are those authorities that have concerns about the impacts of tourism, where key policy issues relate to balancing the needs of locals, visitors and other interests, dealing with impacts arising directly from tourism activity, and managing environmental resources (36% of authorities). A particular concern indicated by three council representatives is that of the cost of developing and managing tourism opportunities, activities and impacts. Two of these indicated impending studies to ascertain the economic cost of infrastructure and attractions, while a third noted the difficulties for councils with small populations to afford infrastructure improvements through the local rates system. 16% of respondents were conversely more concerned about developing tourism assets, promotions and infrastructure in an attempt to generate or meet the demand. Some 40% of respondent did not have any tourism issues of concern. The response may hide a number of more insidious issues, some local agencies do not possess the tourism expertise to identify and deal specifically with tourism impacts, while others may be more focused on championing the marketing orientation of Tourism in generating economic benefits, considering that policy literature focus on poverty reduction strategies. In many cases, there are significant dangers that negative impacts are not anticipated, mitigated or managed. Worth noting though is that 57% respondents identified tourism related issues that needed to be addressed the study. These findings indicate a growing interest and concern about the effects of tourism and the need for local tourism planning authorities to address impacts, both positive and negative, through the planning system. In addition, the range of issues identified, suggesting either a higher level of tourism awareness within councils or the emergence of a more extensive number of impacts.

5.4. Importance of Tourism in Livingstone

As indicated in [16,74] "74.9% of foreign tourists, who have had the opportunity to visit Zambia's popular tourism destinations", visited Livingstone Mosi-oa-Tunya area. Other popular destinations included South Luangwa National Park (24.3%), Mosi-oa-Tunya National Park (25.4%), Lake Kariba (26.5%) and other National Parks and Game reserves (26.5%) [16]. These destinations are Zambia's most developed and marketed attractions [75-78]. Respondents were asked to indicate if the perceived importance of tourism in the case of Livingstone. Some 50% respondents stated that the importance of tourism had increased, 17% of these stating increased significantly. The main reason given for this was the increasing recognition of the realized and potential economic benefits of tourism within the local tourism areas. It appears that many tourism stakeholders have become more aware of the beneficial effects that tourism can bring to a locality through as a source of revenue, business development and employment opportunities. In particular, the awareness of the ability of events to draw visitors to an area appears to have strengthened. Other contributing factors included growth in tourism, improved marketing and strategic vision, development of new products and services, and more central government funding. Only 7% respondents stated that the importance of tourism had decreased, partly due to the limited tourism appeal of one location but in two others a perceived lack of value, for example: Zambia Tourism Board ZTB have been unable to demonstrate, articulate and quantify the value in monetary terms. 26% of the respondents stated that the importance of tourism remained the same. This was explained by several locations where tourism activity remained static or where growth was limited by infrastructure constraints. One issue identified was the absence of effective tourism organisations and regional co-ordination to take tourism developments forward and to illustrate the benefits of tourism to the local communities, thereby not propelling tourism forward as a beneficial economic activity. Development of new attractions and recognition of substantial increases in visitors were cited as the main reasons for the increase in importance. This appears to indicate that tourism area have a clear understanding of how tourism can benefit their locality, which may have resulted from the key messages in the national tourism plan and associated reports. However, similar issues with regard to lack of financial support given to tourism or lack of importance placed on tourism activities.

Table 4. Issues identified by respondents towards sustainable tourism, by author, (2013).

Issues	Number of Responses
Managing adverse environmental effects	3
Need to develop transport infrastructure	3
Waste disposal (especially relating to freedom camping)	3
Weighing up the economic cost of tourism	2
Conflicts between visitors, developers and residents	2
Product development	2
Concerns about effects of specific tourism developments	1
Different approaches adopted by different bodies	1
Addressing seasonality	1
Desire to maintain low impact tourism	1
Increasing demand for outdoor activities and how to meet it	1
Partnership and cultural opportunities	1
Oversupply of road stopping places	1
Effect of climate change on travel patterns	1
Increasing promotions	1
Pressure on infrastructure at peak times	1

5.5. Future Tourism Development

The range and scope of developments (78.6%) indicate a significant rise in the tourism infrastructure across the country, from airport enhancements to visitor trails. The most developments, which had taken place in three local communities in Livingstone (Mosi-oa-Tunya) area followed by accommodation development (non-hotel) in 30 areas (71.4%) [15]. It has been reported that the total number of tourists to Zambia is expected to reach more than 1.4 million tourists in by 2015 and these will require more hotel establishments in the country [15].

Table 5. Tourism planning pressures created in Livingstone greater area, by author (2013).

Tourism Pressures	Number of Respondents
Accommodating more visitors	5
Demands on local services	3
Anti-social behaviour/community spirit	2
Demands on water	2
Effects on wildlife	2
Waste volumes	2
Costs of stopping inappropriate development	1
Ensuring development no environmental degradation	1
Housing affordability for local residents	1
Increased freedom camping	1
Lack of workforce in peak season	1
Need to build more accommodation	1
River/waterfront subdivision	1

The development of new attractions at all levels suggests vibrancy in tourism development. In terms of the types of new developments, the list of new attractions, facilities and services on offer is considerable and far too extensive to include, but incorporates a large proportion of new trails, tours, guided walks and outdoor adventure activities, with a smaller amount of development to create or upgrade cafes, hotels, museums and retailing. All of which utilise environmental resources and all of which have the potential to create or exacerbate adverse impacts. As such, the role of the ZAWA in controlling the effects of tourism development is clear in a climate where growth in individual adventure tourism enterprises and outdoor pursuits is occurring. Some 44.2% of respondents considered local communities in the Livingstone area to be under pressure from increased tourism and Table 5 show the major pressures highlighted by respondents. Three broad categories of responses are distinguishable through examination of a subsequent open question on what pressures existed in localities.

First, specific locations were identified as likely to experience increased visitor numbers and associated impacts, e.g Mukuni's village on the eastern and Inyambo local tourism Community Development Trust areas. Second, the concerns arising from increased visitor numbers were identified including, demand for infrastructure, construction of tourist-related ventures, dealing with municipal waste, water demand and waste water disposal.

Increased freedom camping and effects on wildlife and natural areas, housing affordability, second homes and subsequent loss of community culture attributes, increase in tourist arrivals (e.g. Livingstone's newly extended Harry Mwaanga Nkumbula International airport expansion).

Third, and somewhat in contrast to the latter responses, a grouping of respondents though smaller than the latter, want to grow tourism and maximise the benefits, through creating infrastructure, building more accommodation and increasing the workforce. The survey identified that respondents in areas with the largest number of guest at night of over 5500 in the peak month were more likely to report that their area was under pressure from tourism. Correspondingly those with the smallest number of nights (less than 1000) were the least likely to be under pressure..The areas under pressure tend to include those reliant on the natural environment, cities, areas on the main tourist routes and National Parks. Those not under pressure includes those wishing to develop tourism currently with low visitor numbers and those off the beaten track. Interestingly, 73% local tourism area authorities at (ZAWA, NHCC) respondents (Southern Province region) perceived Livingstone greater areas to be under pressure compared with 29% at central government department at MTENR. Explanations for the perceived higher pressure on the South include respondents' personal experiences in conjunction with often heavy concentrations of packaged tourism and adventure tourism utilising the physical and natural environment.

5.6. Linkages and Synergies Planning (MTENR/ ZAWA/NHCC/ZEMA)

Under Zambian laws consents are required for all tourism developments. Consents are issued by multiple central governmental departments, regional and local authorities and communities depending on the scope of the consent sought [60]. Ascertaining accurate data on tourism related resource consent applications is highly problematic. While many respondents were able to give precise numbers in relation to resource consent applications and refusals, a significant 13 respondents were not able to provide the data. The main reason given for this is that tourism is not always isolated as a key variable in the database recording process for tourism enterprise concession consent applications. Some developments are not primarily designed for tourism purposes but may produce a tourism spin-off, e.g. development of a winery. In other cases, databases are not set up to be readily searched, data is not feed into system as "tourism", but as "commercial activity" and in several cases, the detail of activity or data is not even kept. This seems to indicate an inherent problem in the data management of tourism enterprise concession consent applications with regard to tourism, and a technical inability to retrieve useful information that can inform tourism planning at local, regional and national strategic levels. Acknowledging the limitations of the data, the following results give a broad indication of the workings of the ZAWA process in relation to tourism development within local communities.

Twenty four respondents representing 56% had dealt with tourism enterprise concession consent applications since 2010. The highest number of applications dealt with by one authority was 40. Ten authorities had dealt with between 1 and 10 applications, six between 11 and 20 applications, five between 21 and 30 applications, and three had dealt with 31 or more applications. While the largest number of applications were dealt with by District Councils, participants in the process held the view that all of these tourism programmes are developed by a monoactor form of centralised administration, generally overlooking the knowledge, skills and goals of local tourism organisations, both public and private, 50% of the CBNRM accounted for 37% of tourism enterprise concession which still needed ZAWA approval, indicating a substantial number of applications within a small number of local communities taking the lead role in resource management. Some 76% of tourism concessions submitted were made to the now well established Mukuni Community Development Trust. The trust has established local progressive leadership and used African Wildlife Foundation (AWF) assistance to develop twelve lower level area boards, which is an encouraging result suggesting that local CBNRMs playing a role in receiving enterprise concession related to tourism might have a strategic vision of how tourism should develop in their locality. Importantly, most of the communities receiving large numbers of tourism enterprise concession did have some form of policy guidelines, although two respondents received more than 25 applications did not.

Further, 24% of enterprise concessions were submitted to a start up Sekute Community Development Trust without a tourism guidelines or policy. There is no particular pattern of number of tourism enterprise concession received and the visitor numbers in remote CBNRM areas, with the largest numbers of applications 8 of respondents at ZAWA with over 25 applications in a variety of rural and urban environments, representing those areas that are already important tourism hubs (3 of the 8) and those encouraging the development of a tourism economy (5 of the 8). 3 of respondents at ZAWA received no applications, all of which are in insignificant under developed tourism areas: two not on tourist routes and one within a provincial city environment. One might expect a relationship between those ZAWA provincial offices reporting a large number of tourism enterprise concessions and those reporting that they perceived their area to be under pressure from tourism but this was not the case. 8 ZAWA respondents reported 25 or more applications, 5 respondents stated that their area was not under pressure from increasing tourism. In fact of the 19 of respondents at ZAWA that reported their area to be under pressure, 9 respondents were not able to extract numbers relating to tourism pressure, one ZAWA respondent had never handled tourism enterprise concession applications. A further four respondents received fewer than 10 tourism enterprise concessions, suggesting that it is not necessarily new developments that are creating tourism pressures. Indeed, one might say that applications made under the post [12,50] strategy are perhaps less problematic than existing developments that already generate significant demand.

The notable major challenges identified in the survey include that of poor understanding of what is required in the application for tourism developments activities. Eight respondents 24% of those that had experienced difficulties with applications stated that applications are often presented with incomplete information and a further eight respondents 24% identified lack of understanding and requirements for the tourism establishments under [35,50] process to be a reason why problems are experienced in the application procedure. However, as one respondent commented, early contact with the authorities is important for the process to run smoothly for the applicant: "it is not as bad as they initially think". Similarly, a further difficulty in applications is a lack of consideration of impacts of developments (18%). However, 21% of those that had dealt with ZAWA applications had not experienced any difficulties. As one respondent commented, "ZAWA act is there to protect the environment if a tourism developer follows carefully with ZAWA planning and guidelines/tourism experts, then things appear to go relatively smoothly. Communication between all parties is the key ingredient". The relationship between tourism development, sustainability and the Zambia Tourism Policy and ZAWA act towards tourism, as stated by a respondent: "at the moment the MTENR Tourism Policy of 1999 and ZAWA act of 1998 deals with the sustainability of tourism on a case by case basis, however, at a strategic level the sustainability of tourism is not grappled with, due to outdated policy, legislation and planning". It is also apparent that the Tourism Policy/ZAWA act does not necessarily assure a sustainable approach to tourism planning outside of the particular development under consideration. For example, one respondent noted that: "results from Zambia Tourism Policy and wildlife authority" has been beset with negative administration, transparency and accountability issues". While it is unclear to what extent planning officers work with developers to ensure resource consents are granted, the general premise that there are few outright refusals begs the question as to whether the ZAWA process is rigorous in controlling the negative impacts of tourism in areas under pressure from increased visitor numbers. One respondent commented that "the two institutions MTENR and ZAWA are not a detractor to tourism development", which may or may not be a good thing.

6. IMPLICATIONS

It is clear from the survey findings that the dual role of MTENR/ZAWA in performing a regulatory planning function and promoting tourism raises issues about potential conflicts of interest in applying the Tourism Policy, ZEMA and ZAWA acts while considering the economic development of a locale. This debate is an old one environment versus economics, but in a sustainable development context the need to protect environmental resources to ensure future economic stability is mandatory. It is clear from observations of local communities and authorities at local and regional offices have engaged more actively with the tourism sector through the development of tourism plans and policies. In order to have sustainable development as a national policy direction as reflected in policy developments in the revised sixth national development plans (SNDP: 2011-2013).

6.1. Roles and Integrations of Sustainable Strategies

The integration of Sustainable Development (SD) Strategy in Zambia's National Development Plans' legal framework, various legislation in support of SD developed such as, ZEMA act (2011) to address impacts through strategy preparation is encouraging. However, due inability of local community authorities to benefit directly from the limited resources, especially those with a small population base and limited ability to raise revenue through rates, providing infrastructure, promoting tourism growth and managing impacts are a financial burden on tight budgets from central government: this emerges as a clear theme in the survey. New legislation currently under consideration to minimise waste provides a refund to communities, this is one example of where finding ways to compensate local communities and ratepayers for the use of local services is clearly a challenge and for many councils in Zambia and, indeed, worldwide, juggling the economic costs and benefits of tourism and justifying the outcomes to ratepayers remains problematic.

6.2. Delayed Decentralization Tourism Planning

This study shows that local authorities understand the roles of the MTENR/ZAWA with regard to sustainable tourism, focusing on the effects of tourism activity within their area. "Looking at the bigger picture, one of the criticisms of haphazard sort of implementation due to silo national level planning" [79]. As such, while the intentions of ZAWA in preventing undesirable developments are laudable, the cumulative effects of a number of seemingly innocent, less damaging developments might be equally detrimental. Only one respondent specifically drew attention to this issue, but that does not detract from the importance of the point indeed it might be questioned whether planning officers are sufficiently aware of the dangers posed by this breach within ZAWA framework. Similarly, the focus of ZAWA on effects of activities, while well intentioned, could result in significant economic sectors, like tourism, not been adequately and proactively planned for. Somewhat worryingly, this might be reflected in the lack of response from regional agencies, who do not appear to take tourism as a specific concern under their remit, although are clearly concerned by the effects of tourism such as waste. The inherent difficulties of extracting tourism related projects from MTENR/ZAWA databases held by local tourism authorities appears to be an issue in understanding the implications of the ZAWA for tourism and the extent to which projects are acceptable in the local planning decisionmaking process. Quite clearly, this reflects the inadequacies within data management and retrieval, but also indicates a systemic challenge for the core workings of ZAWA, which by its nature is not concerned with specific industry sectors but with the effects of activities. While the key focus on natural resources provides a valuable framework for the development of appropriate policy and decision-making frameworks, the ability to understand the scope and scale of tourism-related developments is essential particularly given the ambitions of the proposed national tourism strategy (SNDP). Worthy of note is that of the 13 local authorities that were unable to retrieve tourism-related data due to technical problems of record keeping and searching were: four of the eight central government departments (MTENR) stating that their areas were under pressure from increasing tourism; further, two of Zambia's new prime Lusaka circuit very significant local tourist locations; and, further again, three other well-known Livingstone tourism areas. These omissions from knowledge at a planning level indicate the potential to not fully understand the rate of tourism growth from a supply perspective and the cumulative effects of tourism development linked with local aspirations within the confines of long term national development planning (plans) and a budgeting system that is partially decentralised.

7. CONCLUSIONS

The continuing limited involvement of local communities and regional government authorities in tourism planning and development of sustainable tourism approaches existed, given "the continued conspicuously absence of documented national development planning policy and fragmented legislation framework of sustainability in Zambia's national strategies"[60]. With the role of tourism in economic development established and recognised in statutory plans, sustainability now underpins sectorial policy framework for tourism in Zambia, and the landmark steps taken to develop and review national aspirations for tourism development will represent a step forward in establishing a clear remit for local government in planning for tourism. The extent to which this is rhetoric rather than reality is questionable, given the somewhat mixed results in the survey of local government agencies reported in this study. For Zambia a country emerging from a history of centralised economic planning, this question becomes even more vexed quite clearly, a range of pressures continue to affect local areas and the challenges that face many local communities in trying to manage the effects of tourism on environmental resources are as pressing as ever. A national tourism plan will enable local authorities and councils to evolve futures that befit environmental resource opportunities and constraints, community aspirations and local budgets. While tourism is mainly a private sector industry in Zambia, the public sector adopts a dual role as the gatekeeper of tourism developments through planning control, while promoting economic development opportunities through tourism.

As such, while councils have become the arbiters of sustainable tourism through their role in implementing the Zambia Tourism Policy, the appeal of developing the local economy places them in a dichotomous position. While much of this discussion sounds positive, there is still a major gap between strategy and implementation in the evolution towards Zambia as a sustainable destination. While sustainability is now one of the cornerstones under tourism strategy review, much of this lies at a national strategic level and remains as a philosophical stance. Evidence suggests that problems created by tourism pressures do exist and some of these are difficult to deal with given the poor linkages and synergies within the various tiers of government that undertake planning with limited budgets at local government. Pressure at key tourist hotspots and with certain tourism related activities are recognised and with the continuing growth in tourist numbers forecasted, the effects of tourism have the potential to change the nature of the tourist experience and the very foundations on which Zambian tourism is built. The existing problems of geographic concentration of tourism activity will only worsen, exacerbating the pressures on local authorities.

As argued by [64] "policy at a national level that assists local areas in dealing with volumes and the distribution of tourists in a more methodical manner". With reference to [80-84] by enabling more proactive public sector approach to tourism planning, steps towards understanding the dynamics of tourism in Zambia made by the Ministry of Tourism Environment Natural Resource under the Zambia Wildlife Authority by establishing a strategic tourism development model. Given that local government agencies are politically weak, of well-recorded and entrenched patterns of corruption and patronage built around land and planning decisions, this call by planners has a greater degree of cogency as observed by [60,80,85] argue that, "those destinations, localities and nations that prepare to put into practice good detailed policies and strategic plans will reap the benefits for sustaining their tourism products in the future", a cornerstone of Zambia's tourism strategy. Further research and steps would help local Zambian destinations to ensure ZAWA achieves the goals and principles enshrined in the original legislation. Without a more concerted attempt to challenge pro-development policy, Zambia is likely to lose pace in terms of competitive advantage as a clean, green and sustainable tourism destination.

ACKNOWLEDGEMENTS

The authors wish to express their gratitude to Alexandra Thorer & Roy Clarke (kalakikorner) for the assistance in making suggestions, corrections and to the would be referees for their valuable and helpful comments, which have improved the quality of the paper. Also, we extend our thanks to The Copperbelt & Tongji University, especially the Department Landscape studies.

REFERENCES

1. A. Hansen, "The Ecotourism Industry and the Sustainable Tourism Eco-Certification Program (STEP)," International Relations and Pacific Studies, University of California, San Diego, 2013.
2. C. Wu, "Sustainable Development Conceptual Framework in Tourism Industry Context in Taiwan: Resource Based View," Conference of the International Journal of Arts and Science, Vol. 2, No. 1, 2009, pp. 1-11.
3. C. A. Gunn, "Tourism Planning," Routledge, London, 2002.
4. B. Ahn, B. Lee and C. S. Shafer, "Operationalising Sustainability in Regional Tourism Planning: An Application of the Limits of Acceptable Change Framework," Tourism Management, Vol. 23, No. 1, 2002, pp. 1-15.
5. UNEP, "Eco-Tourism in Wider Caribbean Region: AN Assessment," Technical Report No. 31, Caribbean Environment Programme Technical, UNEP, Kingston, 1994.
6. UNEP, "Environmental Codes of Conduct for Tourism," United Nations, New York, 1995.
7. World Tourism Organisation (WTO), "Guide for Local Authorities on Developing Tourism," WTO, Madrid, 1999.
8. C. Tosun, "Challenges of Sustainable Tourism Development in the Developing World: The Case of Turkey," Tourism Management, Vol. 22, No. 3, 2001, pp. 289-303.
9. K. Andriotis, "Tourism Planning and Development in Crete, Recent Tourism Policies and Their Efficacy," Journal of Sustainable Tourism, Vol. 9, No. 4, 2001, pp. 298-3016.
10. V. M. Waligo, J. Clarke and R. Hawkins, "Stakeholder Involvement in the Implementation of Sustainable Tourism," Journal Tourism Management, Vol. 36, 2013, pp. 342-353.
11. UNEP, "The Role of Local Authorities in Sustainable Tourism. Tourism and Local Agenda, 21," United Nations, Rio, 2003.

12. Zambia Wildlife Authority (ZAWA), "Annual Report," Government of the Ministry of Tourism, Environment and Natural Resources, Republic of Zambia Kwacha House, 2007.

13. V. Teye, "Tourism Development in Zambia: Some Physical and Environmental Considerations," Invited Chapter in T. V. Singh and H. L. Theuns, Eds., Towards Appropriate Tourism: The Case of Developing Countries. Peter Lang (European University Studies), Frankfurt, (Revised and Expanded Version of "Geographical Factors Affecting Tourism in Zambia," Annuals of Tourism Research, Vol. 15, No. 4, 1988, pp. 487-509).

14. C. A. Gunn, "Prospects for Tourism Planning: Issues and Concerns A," The Journal of Tourism Studies, Vol. 15, No. 1, 2004, pp. 3-7.

15. Government of the Republic Of Zambia, "Tourism Sector Profile," Zambia Development Agency (ZDA), Lusaka 2013.

16. Zambia Tourism Board, "2010 Visitors' Arrivals Analytical Report," Ministry of Tourism, Environment and Natural Resources, Government of the Republic of Zambia, Kwacha House, 2011.

17. World Bank, "Zambia Economic and Poverty Impact of Nature-Based Tourism," Report No. 43373-ZM, Economic and Sector Work Africa Region 2007.

18. World Tourism Organisation (WTO), "Tourism Highlights 2012" Calle del Capitán Haya, Madrid, 2012.

19. United Nations Environment Programme and World Tourism Organisation, "Tourism in the Green Economy," Background Report, UNWTO, Madrid, 2012.

20. USAID and Sustainable Tourism, "Meeting Development Objectives, USAID's Portfolio in Sustainable Tourism," US Agency for International Development Washington DC, 2005.

21. J. E. Mbaiwa, "Tourism Development, Rural Livelihoods, and Conservation in the Okavango Delta, Botswana, A Doctor of Philosophy Dissertation," The Office of Graduate Studies of Texas A&M University College Station, 2008.

22. W. Jamieson, "Guidelines on Integrated Planning for Sustainable Tourism Development," Economic and Social Commission for Asia and the Pacific, 1999.

23. S. Chiutsi, M. Mukoroverwa, P. Karigambe and B. K. Mudzengi, "The Theory and Practice of Ecotourism in Southern Africa," Journal of Hospitality Management and Tourism, Vol. 2, No. 2, 2011, pp. 14-21.

24. PATA, "Endemic Tourism: A Profitable Industry in a Sustainable Environment," Kings Cross, 1992.

25. WTO, WTTC, Earth Council, "Agenda 21 for the Travel and Tourism Industry: Towards Environmentally Sustainable Development," WTTC, London, 1995.

26. C. Hall, "Historical Antecedents on Sustainable Development: New Labels on Old Bottles?" In: C. M. Hall and A. A. Law, Eds., Sustainable Tourism: A Geographical Perspective, Longman London, 1998, pp. 13-24.

27. UNESCAP, "Managing Sustainable Tourism Development: Tourism Review No. 22," United Nations, New York, 2013.

28. M. Lozano-Oyola, F. J. Blancasa, M. González and R. Caballero, "Sustainable Tourism Indicators as Planning Tools in Cultural Destinations," Journal Ecological Indicators, Vol. 18, 2012, pp. 659-675.

29. International Union for Conservation of Nature (IUCN), "Directory of Afro-Tropical Protected Areas," IUCN, Gland, 1987.

30. S. Metcalfe, "Landscape Conservation and Land Tenure in Zambia: Community Trusts in the Kazungula Heartland," African Wildlife Foundation, Harare Zimbabwe, 2005.

31. V. R. Nyirenda, S. Liwena and H. Kaumba Chaka, "Atlas of the Natonal Parks of Zambia," New Horizon Printing, Lusaka, 2008.

32. Central Statistic Office, "National Accounts Statistical Bulletin," Republic of Zambia 2010.

33. K. T. Taylor and C. T. Banda, "Tourism Development Potential of the Northern Province of Zambia," American Journal of Tourism Management, Vol. 2, No. 1A, 2013, pp. 10-25.

34. Policy Monitoring and Research Centre (PMRC), "Tourism and Wealth Series—Unlocking the Potential of the Tourism Sector to Support Economic Diversification and BroadBased Wealth," Lusaka, 2013.

35. Zambia Wildlife Authority (ZAWA), "Turn-Around Strategy (2011-2016) for the Zambia," 2011.

36. Zambia Tourism Board, "2009 Visitors' Arrivals Analytical Report," Ministry of Tourism, Environment and Natural Resources, Government of the Republic of Zambia, 2010.

37. P. P. Wong, "Coastal Tourism Development in Southeast Asia: Relevance and Lessons for Coastal Zone Management," Ocean & Coastal Management, Vol. 38, No. 2, 1998, pp. 89-109.

38. B. Hatyoka, "UNWTO Offers Great Benefits to Livingstone," Times of Zambia, Business Insights: Essentials. Web. 23 October 2013. UNWTO Offers Great Benefits to Livingstone, 28 August 2013.

39. The Government of the Republic of Zambia, "Zambia National Report," The United Nations Conference on Sustainable Development (Rio+20), Government Printer, Lusaka, 2012.

40. "Reposition Zambia on Tourism Map," Times of Zambia 5 December 2012. Business Insights: Essentials. Web. Retrieved on 23 October 2013.

41. Zambia Wildlife Authority (ZAWA), "The General Management Plan (GMP) for Mosi-oa-Tunya National Park (MoNP)," 1998.

42. Zambia Wildlife Authority, "Five-Year Strategic Plan 2003- 2007," Volume Implementation Plan, Chilanga, 2002.

43. International Union for Conservation of Nature (IUCN), "Biodiversity in Sub-Saharan Africa and Its Islands," IUCN, Gland, 1990.

44. E. Chidumayo, F. Lumbwe, K. Mbata and J. Munyandorero, "Review of Baseline Status of Critical Species Habitats in the Mosi oa Tunya National Park," Department of Biological, University of Zambia, Lusaka, 2003.

45. "Tourism Contributes to Environmental Degradation-President Sata," Lusaka Times, 29 May 2012. Home News: Web. 2012. http://www.lusakatimes.com/2012/05/29/tourism-constributes-environemtal-degradationpresident-sata/

46. B. Liu and F. M. Mwanza, "Questionnaire Website, Sustainable Tourism Development in the Case of Livingstone Zambia," 2012.http://floydmwanza.achievestdzambia.questionpro.com/

47. D. Dredge, "Policy Networks and the Local Organisation of Tourism," Tourism Management, Vol. 27, No. 2, 2006, pp. 269-280.

48. T. Holding, "Environmentally Sustainable Tourism Strategic Plan 2009-2012," Tourism Victoria Victoria, 2009.

49. Government of the Republic of Zambia, "The Tourism Policy for Zambia," Government Printer, Lusaka, 1999.

50. Zambian Environmental Management Agency ZEMA, "Environmental Management Act (EMA)," No. 12. Government Printer, Lusaka, 2011.

51. Policy Monitoring and Research Centre (PMRC), "Tourism Series, Tourism: Supporting Sustainable Development, Income and Job Creation, Zambia," 2013.www.pmrczambia.org.

52. Government of the Republic of Zambia (GRZ), "Fifth National Development Plan: Sustained Economic Growth and Poverty Reduction," Government Printer, Lusaka, 2006.

53. Government of the Republic of Zambia (GRZ), "Sixth National Development Plan: Sustained Economic Growth and Poverty Reduction," Government Printer, Lusaka, 2011, p. 143.

54. Grant Thornton Kessle Feinstein, "Study for Development and Promotion of Tourism in Zambia," 2003.

55. Government of the Republic of Zambia, "Formulation of the National Adaptation Programme of Action on Climate Change (Final Report)," Ministry Of Tourism, Environment and Natural Resources, Lusaka, 2007.

56. Government of the Republic of Zambia, "The National Environmental Action Plan," Ministry of Environment & Natural Resources, Lusaka, 1994.

57. UNEP, "Biodiversity Planning Support Programme, Guide to Best Practices for Sectoral Integration: Integrating Biodiversity into the Tourism Sector," United Nations, New York, 2001.

58. UNEP, "Report on Industry and Sustainable Tourism for the 7th Session of the CSD Tourism and Environment Protection," United Nations, New York, 1999.

59. UNCED, "Agenda 21," United Nations Conference on Environment and Development, Rio de Janeiro, 1992.

60. S. Berrisford, "Revising Spatial Planning Legislation in Zambia: A Case Study," Urban Forum, Vol. 22, No. 3, 2011, pp. 229-245.

61. Zambia Wildlife Authority-ZAWA, "Final Business Plan," Government of the Ministry of Tourism, Environment and Natural Resources, Republic of Zambia, 2006.

62. Zambia Land Alliance (ZLA), "Land Policy Options for Development and Poverty Reduction Civil Society Views for Pro-Poor Land Policies and Laws in Zambia," Lusaka, 2008.

63. K. Andriotis, "A Framework for the Tourism Planning Process," In: A. Raj, Ed., Sustainability, Profitability and Successful Tourism, Kanishka Publishers, New Delhi, 2007.

64. J. Connell, S. Page and T. A. Bentley, "Towards Sustainable Tourism Planning in New Zealand: Monitoring Local Government Planning under the Resource Management Act," Tourism Management, Vol. 30, No. 6, 2009, pp. 867-877.

65. K. Angelevska-Najdeska and G. Rakicevik, "Planning of Sustainable Tourism Development," Procedia-Social and Behavioral Sciences, Vol. 44, 2012, pp. 210-220.

66. K. Kelley, B. Clark, V. Brown and J. Sitzia, "Methodology Matters: Good Practice in the Conduct and Reporting of Survey Research," International Journal for Quality in Health Care, Vol. 15, No. 3, 2003, pp. 261-266.

67. E. R. Babbie, "Survey Research Methods," Wadsworth, Belmont, 1990.

68. J. E. Bartlett II, J. W. Kotrlik and C. C. Higgins, "Organizational Research: Determining Appropriate Sample Size in Survey Research," Information Technology, Learning, and Performance Journal, Vol. 19, No. 1, 2001, pp. 43-50.http://www.osra.org/itlpj/bartlettkotrlikhiggins.pdf.

69. A. Attia, "Planning for Sustainable Tourism Development: An Investigation into Implementing Tourism Policy in the Nwc of Egypt," Thesis Degree of Doctor of Philosophy, Development Planning Unit, The Bartlett School Of Architecture and Planning University College London, University Of London, London, 1999.

70. E. G. Helmy, "Towards Sustainable Tourism Development Planning: The Case of Egypt," A Thesis Submitted in Partial Fulfillment of Doctor of Philosophy, Bournemouth University, Bournemouth, 1999.

71. D. Dierckx, "How to Estimate Your Population and Survey Sample Size," 2013.http://www.check^{m}arket.com/2013/02/how-to-estimate-your-population-and-survey-sample-size/0.

72. "The Poverty Reduction Strategy Paper (PRSP)," Ministry of Finance and National Planning, Government of the Republic of Zambia, 2002.

73. Grant Thornton Associates Limited, "Audit of Zambia Wildlife Authority's (ZAWA's) Current Capabilities and Development of Project Proposals for Institutional Capacity Development over a Five Year Period (2004-2008)," Lusaka, 2004.

74. Government of the Republic of Zambia, "Visitors Arrivals Analyticl Report," Produced By: Ministry of Tourism, Environment and Natural, Lusaka, 2009.

75. World Bank, "Zambia-Support to Economic Expansion and Diversification (SEED): Tourism," Republic of Zambia Ministry of Tourism, Environment and Natural Resources Securing the Environment for Economic Development (SEED) NW, Washington D.C., 2001.

76. Government of the Republic of Zambia, "Mosi-oa-Tunya National Park General Management Plan," ZAWA, Lusaka, 2000.

77. GEF-UNDP-Anchor Environment Consultants, "2001-2006 Economic Analysis and Feasibility Study for Financing Zamibia's Protected Areas," GEF-UNDP-Government of Zambia, 2001.

78. European Development Fund (EDF), "National Parks and Wildlife Service Project: Draft Master Plan," Transtec, Milan, 1998.

79. C. K. Chunga, "Integration of Sustainable Development in Zambia's. National Development Plans," Lessons Learnt &. Recommendations, National Planning Department, Ministry of Finance, 2012.

80. D. L. Edgell, M. D. Allen, G. Smith and J. R. Swanson, "Tourism Policy and Planning: Yesterday, Today and Tomorrow," Butterworth-Heinemann, Oxford, 2008.

81. N. Blaikie, "Designing Social Research, An Introduction to Qualitative Research," Blackwell, Oxford, 2000.

82. P. A. DeGeorges and B. K. Reilly, "Sustainability, the Realities of Community Based Natural Resource Management and Biodiversity Conservation in Sub-Saharan Africa," Economic Growth and Sustainable Wildlife Management, Vol. 1, No. 3, 2009, pp. 734-788.

83. R. Dodds, "Sustainable Tourism: A Hope or a Necessity? The case of Tofino, British Columbia," Canada Journal of Sustainable Development, Vol. 5, No. 5, 2012, pp. 54-64.

84. A. J. Dougill and M. S. Reed, "Framework for Community-Based Rangeland Degradation Assessment for the Kalahari, Botswana," RGS-IBG Annual Conference DARG Session on Sustainable Resource Use: Critical Issues in Developing Areas, The Royal Geographical Society, Kensington Gore, 2003, p. 23.

85. L. Dwyer, D. Edwards, N. Mistilis, C. Roman, N. Scott and C. Cooper, "Megatrends Underpinning Tourism to 2020: Analysis Of Key Drivers For Change," CRC for Sustainable Tourism Pty Ltd., Gold Coast, 2008.

CHAPTER 11

Integrating Cultural and Nostalgia Tourism to Initiate A Quality Tourism Experiences at Chiangkan, Leuy Province, Thailand

A. Nilnoppakun·, K. Ampavat

Faculty of ManagementS Science, Silpakorn University IT Campus, Petchaburi, Thailand

ABSTRACT

This study aimed to examine tourists' demands and Chiangkan tourism resources (supply), and recommended an integrative process to Chiangkan to initiate a Quality Tourism Experiences destination. Tourists' questionnaire was used to collect information from domestic tourists visiting destinations in Thailand (n =700, Chiangkan = 200 and other famous destinations =500). Interviews were employed to tourism services providers (accommodation, transport, and food and beverage) and elites persons (n = 15) in Chiangkan District to gain their opinions towards Chiangkan tourism and its resources. It was found that besides its conservative old houses and local way of life, there are many Buddhist temples in Chiangkan. Additionally, there are many cultural conservatives groups that responsible for 12 Buddhist rituals (for 12 month) each year. Majority of respondents strongly agreed that Chiangkan is a potential destination in terms of social components, attraction, activities, and accessibility. With these quality tourism components that match tourists' demand; Chiangkan is appropriate to develop as a Quality Tourism Experiences destination.

Keywords: Cultural tourism; Nostalgia tourism; Alternative tourism; Quality Tourism Experiences;Tourism components; Tourists' attraction

1. INTRODUCTION

With the tourists' behaviour trend gradually change from conventional "mass" tourism to alternative "niche" tourism, tourist destinations have to improve their tourism products and differentiate the destinations' image to motivate the changing market. The important characteristics of tourism niche markets are that they are likely to have more experiences and knowledge about tourism products and services, and they are more enthusiastic to involve in the tourism process such as reserving their own accommodations, and selecting places to visit (Niezgada, 2013).

To be competitive, destinations should create a Quality Tourism Experiences environment to satisfy tourists. However, the notion of Quality Tourism Experiences is complex and involves many parties and factors during the tourism service process.

2. QUALITY TOURISM EXPERIENCES

According to Model of Tourism Planning and Development suggested by Middleton and Hawkins (1998), tourism experiences are the consequences of the interaction of demand side (Market focus) and supply side (Resource focus). Therefore, the revised model focuses on quality tourism experiences besides tourism resources and activities available at the destination. This focus forces local stakeholders to develop tourism products and images that can satisfy visitors (Pennington-Gray & Carmichael, 2006).

A potential tourist is a person that has ability and motivation to travel. The potential tourists' demands are also preliminary to tourism planning and development (Pearce, 2005). These factors also described in the work of Holloway and Hudson (Holloway, 1994; Hudson, 1999) and many others scholars in terms of ability to travel (spare time, disposable income, property, health, and mobility) and motivation to travel value, attitude, trend, fashion, motive, demands, socio-demography, working and living condition, and life style) (Niezgada, 2013).

Quality tourism experiences notion has been explained in views of the visitor, the tourism developer, the product, the media and the local community that brings together the complication of "quality" concept and its relationship to tourism (Figure 1) (Nickerson, 2006).

Thus, major components of quality tourism experiences can be described in terms of attractions and activities, the social component, and support service components (amenity).

2.1. Attraction and activity

Attraction and activity are always the main factors that tourists choose to describe their travel experiences, often in terms of the quality of attractions and activities experienced (Andereck, et al., 2006). Besides developing of quality attraction (s), activities that can satisfy tourists are also prioritized to tourism planning. Therefore, attraction and activity can be identified as 'pull factors' that draw tourists to a destination.

Tourism attractions include natural attraction such as beach and seaside, waterfall, and national park with flora and fauna; local culture such as local arts and folks, local festival and ceremony, and local way of life; historical sites such as church, temple, and castle; theme parks such as safari park, and Disney World.

Global trend towards 'green' tourism affected tourism activities at destinations. There is a growing tourists' awareness in global warming, hence, green activities such as hiking, trekking, bicycling; and conservation activities such as nurture of sea-life, and beach cleaning are preferred by these groups of tourists.

2.2 The social component

While travelling mediation with family and friend (s) whom tourists travel with is one of important factors affecting tourism experiences (Prentice, et.al. 1998), it consequently influencing the assessment of quality (Andereck, et al., 2006). Additionally, tourists interacting with others, and tourism environments (natural environment and physical environment) are also parts of their experiences. In tourism context, others include other tourists, service providers, public sectors, relating organization, and local communities, who are also called 'stakeholders' (Jennings & Weiler, 2006). Table 1 showed roles of stakeholders and their interaction with tourists.

Even though local communities have informal interaction with tourists, they play an important role as 'host' who receiving 'guests'. In his previous work, Murphy (1985) suggests the necessary of community approach to tourism planning and marketing. Local residents who encouraging tourism tend to be good hosts which resulting in positive tourists' experiences This positive tourists' experiences increases returning guests and their word-ofmouth recommendation to family and friends (Carmichael, 2006).

Murphy and associates (2000) applied a Partial Least Square analysis to test the assumption that destination components such as climate, scenery, ambiance, friendliness to cleanliness influencing quality for visitors to Victoria, BC, British Columbia (beta = 0.36). Hence, they recommend that it is important to tourism industry to focus on the more general environmental components as well as infrastructure and related business to tourism planning and development (Murphy, et al., 2000)

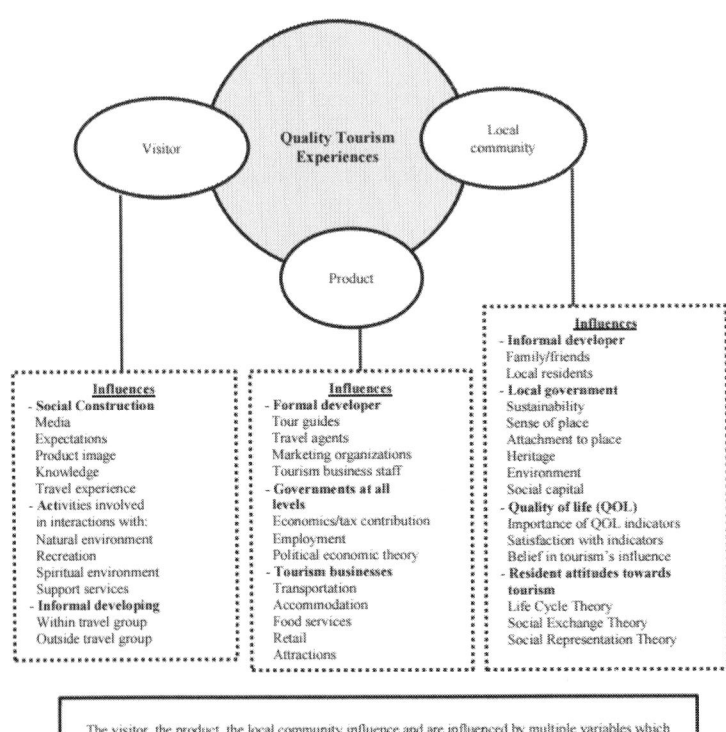

Figure 1. Various Influences on Quality Tourism Experiences Modified from Nickerson, 2006, p.228

Table 1. Roles of stakeholders and their interactions with tourists

Stakeholders	Formal interaction	Informal interaction	Planning and recollection (pre-post visit)	Travel to and from site	On site
Local government	✓		✓		
Travel agent /Tour guide	✓		✓	✓	✓
Transport operator	✓			✓	
Accommodation staff	✓		✓		✓
Food & Beverage provider	✓				✓
Family and friend (s)		✓	✓		✓
Local communities (hosts)		✓			✓
Other tourists		✓	✓		✓
Souvenir /Photo		✓	✓		✓
Non-tourism employee		✓	✓	✓	✓

2.2 Support services component (amenity)

Support services or amenities include transportation, accommodation, food and beverage services, and entertainments. It is necessary that the destinations providing adequate and standard support services to tourists. Infrastructure and public services (hospital, police station) are also considered as support services. Government as developer is responsible in providing infrastructure and public services to residents and visitors (Andereck, et.al, 2006).

Buhalis (2000) also describes five components on supply side that are prerequisites to developing a potential destination in terms of attraction, accessibility, amenity, activity; and ability to develop, manage, and market specific destination.

This study aimed to examine both tourists' demands and tourism resources (supply) at Chiangkan District, Leuy province. Then, initiating an appropriate tourism supply and services (tourism components) that may influence tourists' decision to travel to Chiangkan and proposing a Quality Tourism Experiences framework for the destination.

3. METHODOLOGY

Tourists' questionnaire was used to collect information from domestic tourists (demand side) visiting famous destinations in Thailand such as Samui Island, Pai, and Meung Nan (n=700, Chiangkan= 200 and the other famous destinations=500). Convenience sampling and face-to-face technique were used to tourists who agreed to answer the questions.

Semi-structured interviews were employed to tourism services providers (accommodation, transport, and food and beverage) and elites persons (n=15) in Chiangkan District (supply side) to gain their opinions towards Chiangkan tourism and its resources.

Data was collected and analysed during December 2012 to February 2013.

4. FINDINGS

Chiangkan District is a part of Leuy Province which is located in the northeast of Thailand. With its conservative old houses and local way of life, Chiangkan is recognized among domestic tourists as one of the genuine nostalgia tourism destination.

It was also found that besides its conservative old houses and local way of life, there are many Buddhist temples in Chiangkan. Additionally, there are many cultural conservatives groups that responsible for 12 Buddhist rituals (for 12 month) each year.

4.1 Tourists perceptions towards tourism components

In general, the activity that the majority of respondents who were domestic tourists visiting destinations in Thailand would like to experience the most was arts and cultural events (62.4%).The second activity that the majority of them preferred was the local festival/ceremony (53.6%).

Table 2. Activities that domestic tourists would like to experience at a destination in Thailand (n=700)

Activities that tourists would like to attend	Percentage
Arts and cultural	62.4
Local festival/ceremony	53.6
Environmental conservation activities	45.7
Local way of life	39.9

The majority of respondents travelled with friends (42.3%) and family (37.8%), and the main purpose of visiting Chiangkan were for leisure and tourism (86.5%). Table 3 showed tourists' expectations compared with their experiences during visiting Chiangkan.

Table 3. Tourists' expectations comparing with their experiences during visiting Chiangkan (n = 200)

Chiangkan tourism components	Expectation (%)					Experience (%)				
	5	4	3	2	1	5	4	3	2	1
A conservative destination	16.4	51.2	27.9	4.0	0.5	25.4	44.3	26.8	2.5	1.0
Variety of cultural events and historical sites	16.4	44.8	29.8	9.0	-	19.4	43.3	27.8	8.0	1.5
Maintaining of the local way of life	22.4	40.8	29.8	6.5	0.5	31.8	37.3	22.9	6.0	2.0
Opportunity to engage in cultural events/ceremony	19.9	40.3	31.3	5.5	1.0	29.4	36.3	25.8	5.5	3.0
Variety of natural attractions	16.9	37.8	36.3	8.0	1.0	20.4	35.3	34.8	8.0	1.5

Remark: 1) Expectation score rating from 5 = the highest (expectation) in quality to 1 = the lowest (expectation) in quality

         2) Experience score rating from 5 = the highest quality (experience) to 1 = the lowest quality (experience)

From Table 3, it was evident that the majority of respondents who visiting Chiangkan had highly expectation (score of 4) in quality of every Chiangkan tourism components as follow: a conservative destination (51.2%); destination with variety of cultural events and historical sites (44.8%); maintaining of the local way of life (40.0%); opportunity to engage in cultural events/ceremony (40.3%); and destination with variety of natural attractions (37.8%).

During respondents visitation (experience), it was highlighted that the highest score of 5 (the highest quality) increased in every aspects of their expectation from 3.0 percentage (destination with variety of cultural events and historical sites) to 9.5 percentage (opportunity to engage in cultural events/ceremony). Consequently, from their experiences, the majority of respondents perceived that Chiangkan had high quality in tourism components.

These tourism components can be categorized in terms of attraction, activity, amenity, and social components. Respondents were asked if they agreed with the quality of these Chiangkan tourism components. Table 4 showed tourists' perception towards tourism resources or tourism components in Chiangkan.

Table 4. Tourists' perception towards quality of Chiangkan tourism components (supply) in Chiangkan (n=200)

Chiangkan tourism components	Tourists' Perception (%)					
	5	4	3	2	1	x̄
Attraction						
Chiangkan still has its natural beauty.	22.9	54.2	18.4	2.5	2.0	3.94
Chiangkan residents inherit its local culture and heritage.	29.4	42.7	23.4	3.5	1.0	4.66
Chiangkan residents preserve their local way of life.	28.8	40.3	18.9	9.0	3.0	3.83
Activity						
There are environmentally activities for tourists.	26.4	38.3	27.8	5.5	2.0	3.82
There are bicycle lanes provided for site seeing.	32.8	37.8	22.4	6.0	1.0	3.95
Amenity						
There are adequate signposts, and tourist information centre	23.4	47.3	22.8	4.5	2.0	3.86
Buildings and landscapes unify with its environment	31.4	35.8	17.4	12.4	3.0	3.80
Social component						
Chiangkan residents have hospitality mind.	45.8	32.3	19.9	1.0	1.0	4.21
Tourists had opportunity to join local rites and rituals.	31.3	39.3	20.9	6.0	2.5	3.90
Tourists had opportunity to engage in environmental conservation activities with local community.	32.7	35.4	23.4	7.0	1.5	3.91

Remark: Score rating from 5= highest agreeable to 1= least agreeable

The majority of respondents (45.8%) had highest agreeable (score of 5) that Chiangkan residents have hospitality mind. They were highly agreed (score of 4) with the others social components in terms of tourists had opportunity to join local rites and rituals (39.3%), and tourists had opportunity to engage in environmental conservation activities with local community (35.4%).

In terms of attraction components, the majority of respondents felt highly agreeable (score of 4) that Chiangkan still has its natural beauty (54.2%), the residents inherit the local culture and heritage (42.7%), and the residents preserve their local way of life (40.3%).

The majority of respondents also felt highly agreeable (score of 4) that there are environmentally activities for tourists (38.3%), and there are bicycle lanes provided for site seeing (37.8%).

In terms of amenity provided at the destination, the majority of respondents were highly agreed (score of 4) that there are adequate signposts and information centre in Chiangkan (47.3%), and the building and landscapes unify with its environment (35.8%).

4.2 Opinions of interviewees towards tourism in Chiangkan

The interviewing of tourism services providers (accommodation, transport, and food and beverage) and elites persons (n=15) in Chiangkan District can be described in terms of attraction; accessibility; amenity; activity; social components; and ability to develop, manage, and market specific destination (Buhalis, 2000) couple with its related problems.

4.2.1 Attraction, activity, social components, and ability

It was recommended that public sectors responsible for tourism development and market of the destinations provided more support in promoting Chiangkan cultural events and ceremonies, and local rites and rituals throughout the years, especially; 12 months Buddhist rituals to attract all year tourists visiting to Chiangkan. It was also suggested that local foods should be promoted to attract tourists as well. Other attractions that can pull tourists to Chiangkan included traditional old houses, and 9 historical temples.

Although the walking street is the most famous destination that almost all of the tourists visiting Chiangkan visited for shopping and experiencing nostalgia tourism, and staying overnight at the old houses in the area; it needed local government support in managing space for the better order, and cleanliness.

It was brought to attention that on one hand increasing number of tourists to Chiangkan may benefit economic of the area in terms of revenue earning, on the other hand it may cause negative impact to local economic in terms of higher cost of living. Other negative impacts included environmental degradation, and social mobilization. Therefore, ability of community to manage the destination for sustainable development is required.

It was informed that almost all of tourism stakeholders welcome tourists to Chiangkan and willing to involve in tourism activities being promoted by public and private sectors. In addition, there are many cultural conservatives groups that responsible for 12 Buddhist rituals (for 12 month) each year.

4.2.2 Accessibility

Situated by a river bank, in the former time, Chiangkan main merchandize route was by the river. It was recommended that besides road which is the main mean of transportation to Chiangkan, travelling to Chiangkan or Chiangkan site seeing by the river could the alternative mean of transportation.

5. DISCUSSION AND CONCLUSION

From the findings of this study, it was evident that tourists growing demand in cultural tourism because they would like to experience local arts and culture, and would like to engage in local cultural events and activities. Moreover, tourists perceived Chiangkan as a quality tourism destination in terms of quality of attractions, activities, amenity, and social component, especially, the hosts' hospitality mind that tourists appreciated the most.

On the supply site, local stakeholders recognized of their unique cultural resources that could be develop to attract tourists, besides the nostalgia tourism that is the current image of Chiangkan. However, the local community felt the lack of ability to develop, manage, and market the destination; and needed government support to enhance this tourism component.

With those strengths of quality tourism components that match tourists' demands, it was suggested that Chiangkan should integrate their unique cultural resource to the famous nostalgia tourism to differentiate the tourism attractions in the area. Cultural events and ceremony such as 12 months Buddhist rites and rituals could help increasing number of tourists visiting Chiangkan during the events.

To be Quality Tourism Experiences destination; besides enhancing their ability to develop, manage, and market the destination; local community should create this new image to tourists using variety of media. The era of information technology provides opportunity to low cost e-marketing; hence, promoting the new image of Chiangkan would be effective through website and social media.

The proposed framework is shown in Figure 2.

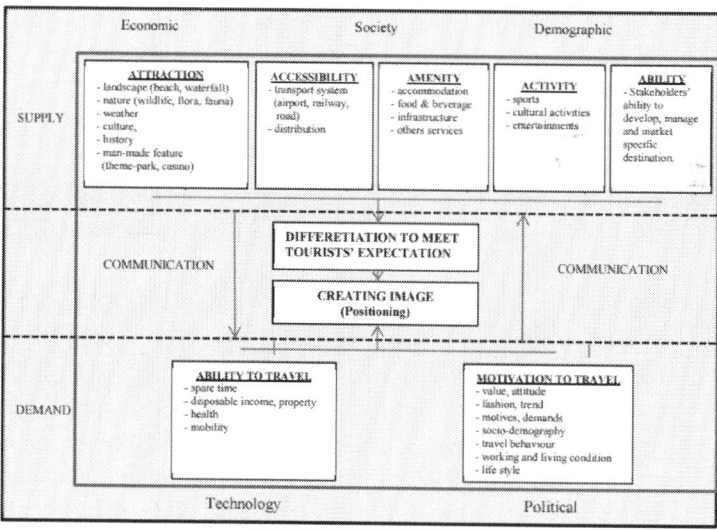

Figure 2. Proposed framework to initiate Quality Tourism Experiences at a destination

It was noted that developing a Quality Tourism Experiences should consider both demand and supply sides. All the tourism components (supply) that influence decision to travel are to be differentiated to match tourists' expectations. It was hope that this proposed framework could be applied to Chiangkan and other tourism destination in Thailand.

REFERENCES

1. Andereck, K., Bricker, K. S., Kerstetter, D., & Nickerson, N. P. (2006). Connecting experiences to quality: Understanding the meanings behind visitors' experiences In G. Jennings & N. P. Nickerson (Eds.), Quality Tourism Experinces. Amsterdam: Elsavier Butterworth Heinemann.

2. Buhalis, D. (2000). Marketing the competitive destination of the future. Tourism Management, 21(1), 97-116.

3. Carmichael, B. A. (2006). Linking Quality Tourism Experiences, residents' quality of life, and quality experiences for tourists In G. Jennings & N. P. Nickerson (Eds.), Quality Tourism Experinces. Amsterdam: Elsavier Butterworth Heinemann.

4. Holloway, J. C. (1994). The Business of tourism. London: Pitman.

5. Hudson, S. (1999). Consumer behavior related to tourism. In A. Pizam & Y. Mansfeld (Eds.), Consumer behavior in travel and tourism. New York: Haworth.

6. Jennings. G & Weiler, B. (2006). Mediating meaning: perspectives on brokering Quality Tourist Experiences. In G. Jennings & N. P. Nickerson (Eds.), Quality Tourism Experiences. Amsterdam: Elsavier Butterworth Heinemann

7. Middleton, V. T. C., & Hawkins, R. (1998). Sustainable tourism: A marketing perspective. Oxford: Butterworth Heinemann.

8. Murphy, P. E, Pritchard, M. P., & Smith. B. (2000). The destination product and its impact on traveller perceptions. Tourism Management. 23, 43-52

9. Murphy, P. E. (1985). Tourism: A community approach. New York: Methuen

10. Nickerson, N. P. (2006). Some reflections on Quality Tourism Experiences. In G. Jennings & N. P. Nickerson (Eds.), Quality Tourism Experiences. Amsterdam: Elsavier Butterworth Heinemann.

11. Niezgada, A. (2013). Prosumers in the tourism market: the characteristics and determinants of their behaviour. Poznan University of Economics Review, 13(4), 130-141.

12. Pearce, P. L. (2005). Tourist behaviour - themes and conceptual schemes. Clevedon: Channel View.

13. Pennington-Gray, L., & Carmichael, B. A. (2006). Political-economic construction of Quality Tourism Experiences. In G. Jennings & N. P. Nickerson (Eds.), Quality Tourism Experiences. Amsterdam: Elsavier Butterworth Heinemann.

14. Prentice, R. C., Witt, S. F.,& Claire, H. (1998). Tourism as experience: The case of heritage park. Annal of Tourism Research. 25, 1-24

CHAPTER 12

Rural Tourism in the South of Spain: An Opportunity for Rural Development

Genoveva Millán Vázquez de la Torre[1], Luis Amador Hidalgo[2], Juan Manuel Arjona Fuentes[1]

[1]Departament of Quantitative Methods, Universidad Loyola Andalucía, Córdoba, Spain

[2]Departament of Economics, Universidad Loyola Andalucía, Córdoba, Spain

ABSTRACT

Tourism is one of the most enriching experiences, and even more so if it involves a rural habitat, in contact with the environment. This kind of tourism, one of the most requested by society currently, offers a great chance for developing rural areas. Rural tourism has become the solution for some problems that have become evident in those areas: the high rate of unemployment, rural exodus and primary sector dependence. So the practice of this activity will generate and diversify income, and create employment. For this reason it is necessary to offer a product adapted to consumer tourist demand, and therefore it is essential to know profile of the tourist. The Andalusian region, in the south of Spain, is famous for being a tourist area of great singularity exhibiting different degrees of development and models of touristic exploitation, strongly characterized by its offer of sun and beach. Today, this community does not limit itself to offering only a sun and beach experience since not all tourists that choose it as their destination have exclusive preference for this sector. Among all the new proposals, one modality is rural tourism, which is the focal point of this study, especially those initiatives for rural tourism in natural parks. Rural tourism is a development factor that will help to correct regional imbalances. Developmental politics related directly to wealth generation in these rural areas can be bolstered by this activity. This study will estimate a forecast to model monthly rural tourist demand in Andalusia for 2014. Its aim is to reveal its evolution in the immediate future in order to propose measures to encourage activity in said tourism sector, by making use of the 24 natural parks and protected areas located in this region.

Keywords: Rural Tourism; Rural Development; Natural parks; Andalusia; Demand Forecast

1. INTRODUCTION

Rural tourism is a tool to further regional development where there is a socio-economic imbalance. In Spain, rural areas with low levels of income and productivity still prevail. Generally, they focus their production on economic activities directly related to the primary sector and suffer high rates of unemployment. They need to diversify their income.

These regions need sustainable growth through an economic culture based on the efficient administration of rural resources, involving the population whose main aim is to achieve a socio-economic and environmental balance. According to Etxano [1], rural tourism will provide additional income to what they already receive and stable employment. All in all, it will contribute to reducing poverty and redistributing income. We do not propose making tourism the main source of income in these regions, but we feel it could be an additional income contribution for the inhabitant.

Mediano [2] and Buckley [3] point out that when tourists wish to enjoy the natural environment, they are inclined to opt for rural tourism in contact with that natural environment to get away from traditional sun and sea tourism, because the latter tourism no longer satisfies new inclinations that have been appearing in our society in recent years. What makes rural tourism a more popular emerging tourism is that the environment is a cause for concern in society, involving the search for sustainable development along with the need to seek satisfaction in one's free time, for example by being in contact with nature or staying with a family. It satisfies needs that the tourist feels nowadays to a greater degree, assert Devesa, Laguna & Palacios [4]. A new market niche has made its appearance due to this change in values of current tourist consumers. Rural areas can take advantage of this niche to generate additional income.

2. RURAL TOURISM: ECONOMIC FACTOR IN RURAL AREAS

Although our research deals with rural tourism, we are not going to delve too deeply into the complex current bibliography to define what rural tourism means and involves. Experts use different criteria to establish their definitions with clarity and accuracy which means that even today there is no unanimous one. In fact, we can find that the term has different meanings. The first one, by Traverso

[5] says: "the tourist activity of sustainable establishment in the rural environment" and the second, by Blanco [6] says: "a singular expression of the new ways of tourism, characterized by: 1) being developed away from urban settlements generally; 2) using diverse natural or cultural resources characteristic of rural environment resources; 3) contributing to local development and to tourism diversification and competitiveness.

Consequently, as Lane [7] says, rural tourism in its purest form should be:

- Located in rural areas.

- Functionally rural—built upon the rural world's special features of small-scale enterprise, open space, contact with nature and the natural world, heritage, traditional societies and traditional practises.

- Rural in scale—both in terms of buildings and settlements—and, therefore, usually small-scale.

- Traditional in character, growing slowly and organically, and involved with local families. It will often be very largely controlled locally and developed for the long term good of the area.

- Of many different kinds, representing the complex pattern of rural environment, economy, history and location.

Starting from these premises, Sanagustín, Moseñe and Gómez [8] consider that rural tourism will help to make the rural area more dynamic and up-to-date, helping to solve arduous socio-economic problems that these regions have to face up to. Dependence on generated income in the primary sector is also one of those problems as seen in Figure 1.

Rural Andalusia has a large population working in the agricultural sector. Statistical studies show that in over 50% of Andalusian towns, there are more than 25% of the population working in the primary sector, thus demonstrating an urgent need to diversify income. Rural tourism represents a key component in the socio-economic development in the region, making it possible to diversify the income of the rural population to guarantee prosperous development of the zone as it proposes an activity that will generate additional income and an element that distributes that income. Soteriades considers [9] indoor areas have to be taken into consideration to make the most of the opportunity that has emerged. They can fill a gap in the market because of the new interest in this type of tourism in recent decades with respect to sportive, cultural, and gastronomic activities in a rural environment.

Multiple factors have had a bearing on the change in tourist consumer's habits and values; it is essential to define the profile of this tourism consumer to offer a specialised product adapted to the needs of this tourism.

According to Millán & Melian [10], rural tourism is an economic factor with some features that make it special. This tourism is carried out with more periodicity, decreasing the typical seasonal nature of these zones which is of great importance from an economic point of view, mainly because the employment created will not be temporary like most employment created by tourism. Tourism leads to a multiplying effect in the population of these rural areas. It diversifies the income obtained; it decreases dependence on the primary sector and reduces the economic risk which characterises it. Parada & Rodriguez assert [11] that rural tourism will help to redistribute income between regions and will encourage the creation of employment directly.

At the same time, Ciruela notes [12] that the establishment of tourist activity in rural areas may be an element that depends on other areas. It is an exodus from urban areas on the part of the population. Equally, this tourism will help to decrease the high rates of unemployment. These two are the greatest difficulties facing these developing societies.

Fuller states [13]: it is not a question of making tourism the main income resource in these areas; on the contrary, it is a contribution of additional income for inhabitants because if the opposite took place, it would entail a high probability of saturation in the rural area. Miller et al. [14] proposes the promotion of a sustainable environment to generate wealth and employment in the course of time. So, maintenance and conservation of the environment will lead to a tourist development that is respectful of the natural and cultural environment.

What are equally essential are co-operation, dialogue and the co-ordination of socio-economic factors and institutions (local, regional and supranational). Furthermore, it is necessary to implicate society and to educate it. Of course, the participation of the people is a momentous factor in raising the level of association between towns. And one of the main weaknesses is the increasing public awareness of this potential boon to be exploited in the region. It is necessary to raise the region's inhabitants' degree of commitment toward tourism matters, because they will be the most benefited. Equally, public bodies should promote development plans to make tourism activity easier in these zones. Although the growth of every region has to come from within the region itself, we should consider the resources and possibilities of the area to encourage them to offer a product which satisfies tourist consumers' needs fully.

Anton, Ne-lo & Orellana [15] argue that tourist activity will not cancel out that of agriculture but will help to correct regional imbalances. Small growers and family farms will be able to benefit from the greater involvement of society in this sector. But, according to Buckey [16] and McAreavey &McDonagh [17], rural tourism will also spark off negative effects in these areas; in the long-term, the cost of living will increase, agricultural area will decrease because of all these things; it will be necessary to have effective and efficient planning and management of all regional resources and be aware that tourist destinations should not run the risk of flooding, which would cause them to lose all their charm, affirm Zhong, Deng, Song and Ding [18].

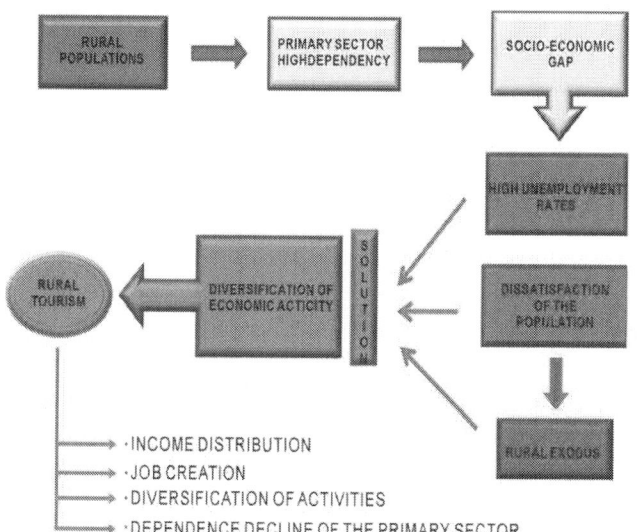

Figure 1. Rural tourism concept map.

3. RURAL TOURISM IN ANDALUSIA

3.1. Introduction

The regional autonomy of Andalusia exhibits a precarious economic situation with high unemployment rates (36.37%), 10 points over the Spanish average (28.98%), along with low activity rates (58.94% compared to the 59.94% national average for the third trimester of 2013). It is one of the communities most affected by the economic crisis, which is especially serious in the rural areas of the region.

Analyzing employment by sectors (Table 1), the Andalusian population comprises 2.53 million people, which represents 15% of those employed nationwide, By sectors, 6.1% of workers are dedicated to agriculture, which is 2 points higher than the Spanish average of 4.2, and 79.7% are in the service sector, especially highlighting those whose professional activity is dedicated to tourism (restaurants, hotels and shops).

On the other hand, the region has many important natural resources (24 natural parks and protected zones, Figure 2) and cultural heritage located in rural areas. In these natural spaces it is possible to develop an activity like tourism to help generate complementary economic income for those in the agricultural sector.

Since these natural resources can be used in a sutainable way, public administrations have lent their support to tourism as an opportunity to transform a rural productive structure, already obsolete as a result of the loss of interest in its facet as a supplier of primary products. In this sense, community initiative through the application of Leader structural funds (from the European Union), and the Spanish program PRODER (set of rural development programs that incorporate endogenous development initiatives that have been implemented exclusively in Spain) have shown to be efficient instruments, although it is true that, according to Pulido and Cardenas [20], there have been diverse results in different regions of Spain.

Table 1. Employment in each economic sector (third trimester 2013).

Sectors	Andalusia (thousands of persons)	% Total Andalusia	Spain (thousands of persons)	% Total Spain
Agriculture	155.6	6.1	705.6	4.2
Industry	217.9	8.6	2280.2	13.5
Construction	139.3	5.5	1013.5	6.1
Services	2019.5	79.7	12823.8	76.2
TOTAL	2532.3	100	16823.1	100

Data source: National Institute of Statistics [19].

Figure 2. Natural Parks in Andalusia.

3.2. Evolution of Rural Tourism

In general, since the outbreak of the financial crisis, the Andalusian tourism sector has gone downhill, An example is the volume of employment created in this sector. In 2008 Andalusia held the first place in the ranking of regions (16.4% of all jobs generated in the tourism sector at national level), falling to third position in 2010 (15.3% of all jobs generated in the tourism sector at national level). This has entailed, among other consequences, the destruction of 33.500 jobs in the period in question. Rural Andalusian tourism, in particular, has also been negatively affected by the adverse economic situation. Figure 3 shows the evolution of rural tourism in the 8 Andalusian provinces in 2000-2010. Almost all the provinces have undergone a decrease in their number of rural tourists since 2009 as a consequence of the economic crisis, since most visitors come from the Andalusian community itself or other Spanish regions.

4. METHODOLOGY

4.1. ARIMA Models

A time series is defined as a set of data collected sequentially in time. It has the property that neighbouring values are correlated. This tendency is called autocorrelation. A time series is said to be stationary if it has a constant mean and variance. Moreover the autocorrelation is a function of the lag separating the correlated values and called the autocorrelation function (ACF).

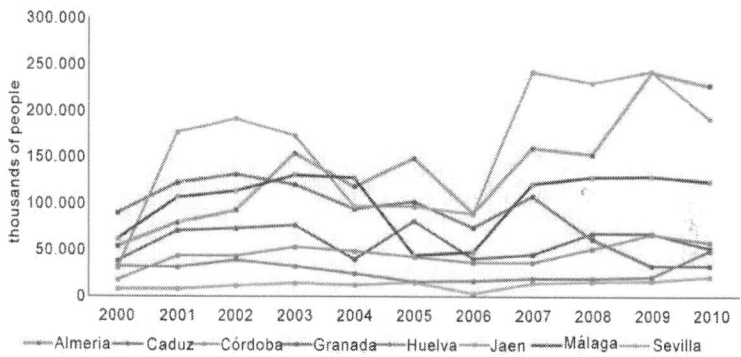

Figure 3. Evolution of rural tourism in Andalusian provinces 2000-2010.

A stationary time series $\{Y_t\}$ is said to follow an autoregressive moving average model of orders p and q (designated ARMA(p, q)) if it satisfies the following difference equation:

$$Y_t - \phi_1 Y_{t-1} - \phi_2 Y_{t-2} - \phi_3 Y_{t-3} - \cdots - \phi_p Y_{t-p}$$
$$= a_t - \theta_1 a_{t-1} - \theta_2 a_{t-2} - \theta_3 a_{t-3} - \cdots - \theta_q a_{t-q} \tag{1}$$

or:

$$\phi(B)Y_t = \theta(B)a_t \tag{2}$$

where $\{a_t\}$ is a sequence of uncorrelated random variables with zero mean and constant variance, called a white noise process, and the ϕ_i's and θ_j's constants;

$$\phi(B) = 1 - \phi_1 B^1 - \phi_2 B^2 - \phi_3 B^3 - \cdots - \phi_p B^p$$
$$\text{and } \theta(B) = 1 - \theta_1 B^1 - \theta_2 B^2 - \theta_3 B^3 - \cdots - \theta_q B^q$$

and B the backward shift operator defined by:

$$B^k Y_t = Y_{t-k}$$

If $p = 0$, the model (1) becomes a moving average model of order q (designated MA(q)). If, however, q = 0 it becomes an autoregressive process of order p (designated AR(p)). An AR(p) model may be defined as a model whereby a current value of the time series Y_t depends on the immediate past p values: $Y_{t-1}, Y_{t-2}, Y_{t-3}, \cdots Y_{t-p}$. On the other hand an MA(q) model is such that the current value X_t is a linear combination of immediate past values of the white noise process: $a_{t-1}, a_{t-2}, a_{t-3}, \cdots, a_{t-q}$. Apart from stationarity, invertibility is another important requirement for a time series. It refers to the property whereby the covariance structure of the series is unique. Moreover it allows for meaningful association of current events with past history of the series.

An AR(p) model may be more specifically written as:

$$Y_t - \phi_{p_1} Y_{t-1} - \phi_{p_2} Y_{t-2} - \phi_{p_3} Y_{t-3} - \cdots - \phi_{p_p} Y_{t-p} = a_t$$

Then the sequence of the last coefficients $\{\phi_{ii}\}$ is called the partial autocorrelation function (PACF) of $\{Y_t\}$.

The ACF of an MA(q) model cuts off after lag q whereas that of an AR(p) model is a combination of sinusoidals dying off slowly. On the other hand the PACF of an MA(q) model dies off slowly whereas that of an AR(p) model cuts off after lag p. AR and MA models are known to exhibit some duality relationships. These include:

1) A finite order AR model is equivalent to an infinite order MA model.

2) A finite order MA model is equivalent to an infinite order AR model.

3) The ACF of an AR model exhibits the same behaviour as the PACF of an MA model.

4) The PACF of an AR model exhibits the same behaviour as the ACF of an MA model.

5) An AR model is always invertible but is stationary if $\phi(B)=0$ has zeros outside the unit circle.

6) An MA model is always stationary but is invertible if $\theta(B)=0$ has zeros outside the unit circle.

Parametric parsimony consideration in model building entails preference for the mixed ARMA fit to either the pure AR or the pure MA fit. Stationarity and invertibility conditions for model (1) or (2) are that the equations $\phi(B)=0$ and $\theta(B)=0$ should have roots outside the unit circle respectively.

Often, in practice, a time series is non-stationary. Box and Jenkins [21] proposed that differencing of appropriate order could render a non-stationary series $\{Y_t\}$ stationary. Let degree of differencing necessary for stationarity be d. Such a series $\{Y_t\}$ may be modeled as:

$$\phi(B)\nabla^d Y_t = \theta(B)a_t$$

(3)

where $\nabla^d = (1-B)^d$ and in which case $\phi(B)\nabla^d = 0$ shall have unit roots d times. Then differencing to degree d renders the series stationary. The model (3) is said to be an autoregressive integrated moving average model of orders p, d and q and designated ARIMA(p, d, q).

4.2. Seasonal ARIMA Models

A time series is said to be seasonal of order d if there exists a tendency for the series to exhibit periodic behaviour after every time interval d. Traditional time series methods involve the identification, unscrambling and estimation of the traditional components: secular trend, seasonal component, cyclical component and the irregular movement. For forecasting purpose, they are reintegrated. Such techniques could be quite misleading.

The time series $\{Y_t\}$ is said to follow a multiplicative (p, d, q) × (P, D, Q)_s seasonal ARIMA model if:

$$\phi(B)\Phi(B^s)\nabla^d\nabla_s^D Y_t = \theta(B)\Theta(B^s)a_t$$

(4)

where Φ and Q are polynomials of order P and Q respectively. That is:

$$\Phi(B^s)=1-\Phi_1 B^{1s}-\Phi_2 B^{2s}-\cdots-\Phi_p B^{Ps}$$

(5)

$$\Theta(B^s)=1-\Theta_1 B^{1s}-\Theta_2 B^{2s}-\cdots-\Theta_Q B^{Qs}$$

(6)

where the Φ_i and the Q_j are constants such that the zeros of the equations (5) and (6) are all outside the unit circle for stationarity or invertibility respectively. Equation (5) represents the autoregressive operator whereas equation (6) represents the moving average operator.

Existence of a seasonal nature is often evident from the time plot. Moreover for a seasonal series the ACF or correlogram exhibits a spike at the seasonal lag. Box and Jenkins [21] and Madsen [22] are a few authors that have written extensively on such models. A knowledge of the theoretical properties of the models provides basis for their identification and estimation, assert Harrison [23].

5. THE DEMAND FOR RURAL TOURISM IN ANDALUSIA

5.1. The Model of Demand

The demand for rural tourism in Andalusia is a seasonal variable; the month of August receives the greatest influx of tourists in rural areas, coinciding with summer vacations (Figure 4). During this period, certain rural areas find themselves saturated with visitors and unable to adequately meet existing demand since there is too little hotel accommodation, and local businessmen are not interested in widening the offer since there is little occupancy the rest of the year, except for long weekends and holidays.

This section has tried to model the rural tourist variable in Andalusia, taking a sample that goes from January 2000 to December 2010 (132 data). To do so, the BoxJenkins methodology (BJ) is what is used, technically known as the ARIMA methodology. According to Gujarati [24], the virtuality of this forecasting method is not in the construction of uniequational models or simultaneous equations, but in the analysis of probabilistic or stochastic properties of the economic time series itself (in this case of the number of rural tourists in Andalusia), under the philosophy of "allowing the information to speak for itself". Unlike regression models where the rural tourist variable can be explained by other regressors like consumer price index, income, etc., in the time series models (BJ) the rural tourist variable can be explained by past or lagging values and by stochastic terms of error. The model used is a seasonal univariate ARIMA:

$$\phi(B)\Phi\left(B^s\right)(1-B)^d\left(1-B^s\right)^D Y_t^{(\lambda)} = \theta(B)\Theta\left(B^s\right)a_t$$

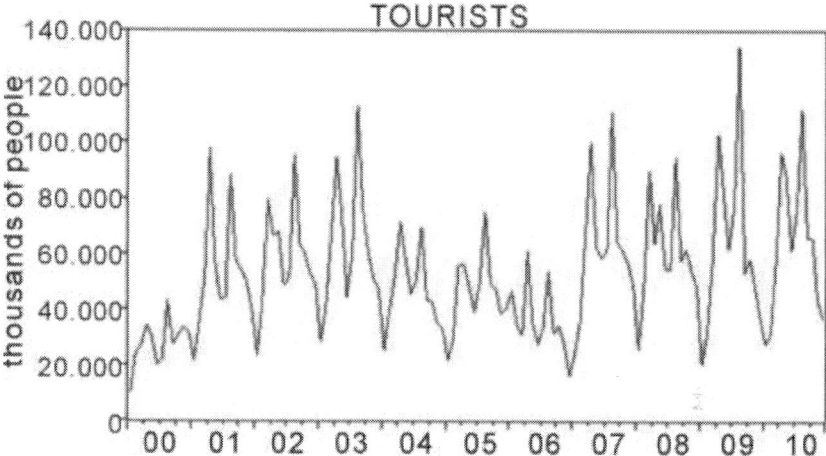

Figure 4. Evolution of rural tourism in Andalusia month by month (January 2000 to December 2010).

where the parameter λ represents the correction of the tendency of variance of the series, and the nabla polynomials $\left((1-B)^d(1-B^s)^D\right)$ represent the correction of tendencies in average and cycle. The model obtained has been validated, both in its coefficients as well as in errors, through the Dickey-Fuller's unit root statistics and the ARCH test.

5.2. Results: Estimating the Demand for Rural Tourism in Andalucia for 2014

The demand for rural tourism in Andalusia, called tourism in the model, is a variable with variance tendency, which has been corrected with the Box-Cox transformation Cox λ = 0.3 (tourism^0.3), and tendency in average and cycle that has been corrected with average and cyclical differentiation. The estimated model of monthly demand prediction for rural tourism in Andalusia is:

$$\left(1+0.485303B^{12}\right)\left(1-B\right)^1\left(1-B^{12}\right)^1 \text{tourism}^{0.3}$$

$$t_{\Phi_1=5.825*}$$

$$=\left(1+0.581937B\right)a_t$$

$$t_{\theta_1=7.267*}$$

The Table 2 includes data of the ARCH test and the Dickey Fuller unit root test, which show that the model has no autocorrelation and is stationary.

This model has been used to make monthly predictions for the year 2014, which can be found inTable 3 and Figure 5. The deduction is that the evolution of tourism will follow similar tendencies to 2013, revealing accentuated seasonality in the month of August. With respect to the evolution of rural tourism figures, between 2010 and 2014 there is a decrease in the total volume of visitors (from 760.092 to 757.126).

Table 2. ARCH test and unit root test.

Heteroskedasticity Test: ARCH		
F-statistic	3.102654	Prob.F(1.104) 0.0811
Obs*R-squared	3.070711	Prob. Chi-Square(1) 0.0797

*Significant parameters $\alpha = 0.05$;
Absence of conditional autoregressive heteroskedasticity.

Null: unit root (assumes individual unit root process)				
Im, Pesaran and Shin W-stat	−12.1768	0.0000	1	117
ADF - Fisher Chi-square	76.3364	0.0000	1	117
PP - Fisher Chi-square	83.2745	0.0000	1	118

Stationary series.

Table 3. Forecast for rural tourism demand in Andalusia for 2014 (thousands of persons).

	Year 2010	Year 2014
January	27.722	26.952
February	31.091	30.128
March	56.603	54.248
April	96.199	98.125
May	88.145	85.369
June	61.214	62.402
July	76.193	76.095
August	111.884	112.324
September	65.991	64.247
October	65.431	66.789
November	42.949	44.658
December	36.670	35.789
TOTAL	760.092	757.126

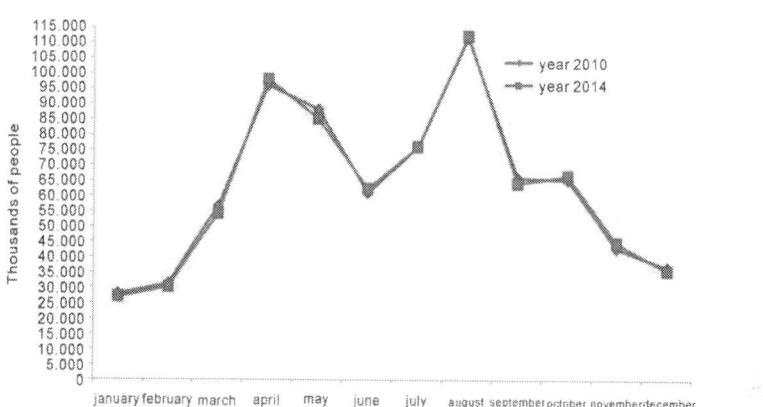

Figure 5. Monthly forecast for the number of rural tourists in Andalusia for 2014 and compared to 2010.

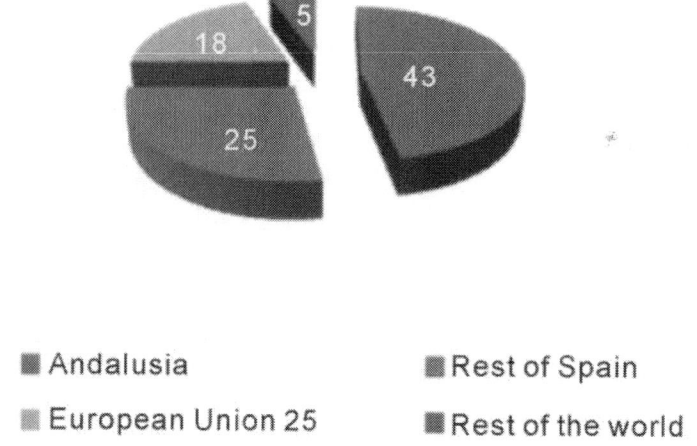

■ Andalusia   ■ Rest of Spain

■ European Union 25 ■ Rest of the world

Figure 6. Place of origin of rural tourism visitors in Andalusia.

In order to alleviate this fall in tourism and the marked seasonality of visits, new markets are proposed. To do so, it would be necessary to promote the opportunities that these rural areas offer in international tourism fairs in order to attract foreign tourists. It must be kept in mind that currently 77% of those visiting these areas are Spanish (43% from Andalusia and 34% from the rest of Spain) (Figure 6).

Therefore, a solution at the current juncture of rural Andalusian tourism would be to attract tourists from other countries where the crisis has not been so serious or is already in phase of recuperation. Moreover, the offer has to be oriented toward those segments of the population with higher buying power, which would also lead to more overnight stays and higher average spending.

6. CONCLUSIONS

In conclusion, we will briefly sum up what has been discussed up to now. The first section sets out how a rural area has to develop into a tourist product. Thus, this sector can provide additional income sources for inhabitants of the region. Also, with the help of public and private bodies fomenting all potential attractions, the rural area can consolidate itself as a generator of a great part of regional economy and tourist destinations. These areas have to adopt measures to improve the image of the area, to promote a type of tourism where visitors can get involved and share in the every-day life of their destinations (daily tasks, customs, ways of local life…), simultaneously rescuing folk activities. All these will offer a degree of differentiation with respect to other rural destinations in the products and services offered by the area.

Although markets for rural tourist destinations are progressing at an incredible rate, in Andalusia we are still in time to be the first region in supply rural tourism. Therefore companies, not only hotels but all the complementary services (area stores, craftsmen…), should anticipate with swiftness and precision, taking into consideration the forecast for demand which can help them to invest appropriately in this wealth resource which is rural tourism. However it is also true that there has to be a change in tourism promotion policies in Andalusia, which have always been too oriented towards the beach and sun.

Rural Andalusian tourism has experienced a drop in the number of visitors, to a great extent, due to its mostly national tourism from Spanish regions that have been greatly affected by the financial crisis. In this sense, the origin of tourists must be diversified to include the whole international market.

Natural Parks are natural areas, little transformed by human exploitation or occupation that, due to its beautiful landscapes, the representativeness of ecosystems or the uniqueness of its flora, fauna or geomorphological formations, have ecological values, aesthetic, educational and scientific conservation deserve preferential treatment. There are 24 national parks and protected areas in Andalusia which can help to increase the number of visitors, and reverse this falling trend in demand in the forecast made by the ARIMA model. It is noteworthy that the special contribution to development and economic diversification of these natural spaces are given by the emergence of new tourist activities like saltworks tourism, birdwatching and agritourism. These activities are configured as future economic activities, with the potential to meet social, economic and environmental functions such as setting the population and conservation of natural and cultural heritage of these territories. They usually develop in slightly modified natural environments and present an offer of nature, adventure, culture and tradition. These tourism activities are managed by the local community, which can be integrated into environmental conservation activities or rural work and also have the potential to serve as an awareness tool for residents and visitors, and as a factor of local socio-economic diversification.

However, companies that are located in these natural areas have to face certain problems. They are usually small family businesses managed in a traditional way, and the staff training is low. These characteristics give them a great difficulty of innovation in products and processes, a low transfer of good practices and knowledge between them and a considerable lack of channels to reach their customers.

The solution for businesses located in these natural parks involves the assumption that innovation is not just developing new products, services, processes and technologies. Instead, what is important is innovation in models that allow the development of the brand "Natural Parks of Andalusia".

The design of a real tourism policy for these spaces should be the framework in which to insert any tourist activity that can generate economic expansion. Regional authorities and businesses in the sector should modify their current policy of tourism promotion in the region to stress the importance of interior tourism and create a framework for rural Andalusian tourism so that this region is no longer known exclusively for its sun and beach alternatives, but also as a region with a rural milieu that is an as-yet-hidden treasure.

Finally, the inhabitants of these rural Andalusion zones have a real need to increase and diversify their sources of income. To do so, rural tourism is an optimum option to help alleviate the long-suffering economy of the inhabitants in these areas.

REFERENCES

1. I. Etxano, "Desarrollo rural en Espacios Naturales Protegidos: el caso del Parque Natural de Gorbeia (1994- 2008)," Lurralde: Investigación y Espacio, Vol. 32, 2009, pp. 197-226.

2. L. Mediano Serrano, "La Gestión de Marketing en el Turismo Rural," Ediciones Pearson Prentice Hall, Madrid, 2004.

3. R. C. Buckley, "Tourism and Environment," Annual Review of Environment and Resources, Vol. 36. 2011, pp. 397-416.

4. M. Devesa, M. Laguna and A. Palacios, "The Role of Motivation in Visitor Satisfaction: Empirical Evidence in Rural Tourism," Tourism Management, Vol. 31, No. 4, 2010, pp. 547-552.

5. J. Traverso Cortés, "Comunicación Interpretativa: Variable Clave en el Marketing-Mix de las Empresas de Turismo Rural," Estudios Turísticos, No. 130, 1996, pp. 37- 50.

6.   R. Blanco and J. Benayas, "El Turismo Como Motor del Desarrollo Rural. Análisis de los Proyectos Subvencionados por Leader I," Revista de Estudios Agrosociales, No. 169, 1994, pp. 119-147.

7.   B. Lane, "What Is Rural Tourism?" Journal of Sustainable Tourism, Vol. 2, No. 1-2, 1994, pp. 7-21.

8.   M. V. Sanagustín Fons, J. A. Moseñe Fierro and M. Gómez y Patiño, "Rural Tourism: A Sustainable Alternative," Applied Energy, Vol. 88, No. 2, 2011, pp. 551-557.

9.   M. Soteriades, "Clusters et Reseaux Dans le Cadre du Tourisme Rural: L'experience Grecque," Revista de la SEECI, No. 23, 2010, pp. 85-117.

10.   G. Millán and A. Melian, "El Turismo Rural en el sur de España: Análisis de la Oferta y Demanda," CULTUR: Revista de Cultura e Turismo, No. 2, 2010, pp. 69-91.

11.   A. F. Parada and J. Rodríguez, "Economic Valuation of Parque Nacional El Gúacharo, Monagas state, Venezuela," Revista Científica UDO Agrícola, Vol. 8, No. 1, 2008, pp. 88-97.

12.   A. M. Ciruela, "Diversificación de la Actividad Agraria Hacia el Turismo Rural: Un Modelo de Decisión Basado en Sociedades Cooperativas Agrarias Oleícolas," CIRIEC: España. Revista de Economía Pública, Social y Cooperativa, No. 61, 2008, pp. 205-232.

13.   N. Fuller, "Reflexiones Sobre el Turismo Rural Como via de Desarrollo: El Caso de la Comunidad de Antioquía, Perú," Estudios y perspectivas en turismo, Vol. 20, No. 4, 2011, pp. 929-942.

14.   G. Miller, K. Rathouse, C. Scarles, K. Holmes and J. Tribe, "Public Understanding of Sustainable Tourism". Annals of Tourism Research, Vol. 37, No. 3, 2010, pp. 627-645.

15.   S. Antón, M. Nel-lo and A. Orellana, "Coastal Tourism in Natural Parks. An Analysis of Demand Profiles and Recreational Uses in Coastal Protected Natural Areas," Revista Turismo & Desenvolvimento, Vol. 7-8, 2007, pp. 69- 81.

16.   R. Buckey, "Sustainable Tourism: Research and Reality", Annals of Tourism Research, Vol. 39, No. 2, 2012, pp. 528-546.

17.   R. McAreavey and J. McDonagh, "Sustainable Rural Tourism: Lessons for Rural Development", Sociologia Ruralis, Vol. 51, No. 2, 2011, pp. 175-194

18.   L. Zhong, J. Deng, Z. Song and P. Ding, "Research on Environmental Impacts of Tourism in China: Progress and Prospect," Journal of Environmental Management, Vol. 92, No. 11, 2011, pp. 2972-2983.

19.   Instituto Nacional de Estadística, "Anuario Estadístico de España 2013," INE, Madrid, 2013.

20. J. I. Pulido and P. J. Cárdenas, "El Turismo Rural en España: Orientaciones Estratégicas Para una Tipología aún en Desarrollo," Boletín de la Asociación de Geó- grafos Españoles, No. 56, 2011, pp. 155-176.

21. G. E. P. Box and G. M. Jenkins, "Time Series Analysis, Forecasting and Control," Holden-Day, San Francisco, 1976.

22. H. Madsen, "Time Series Analysis," Chapman & Hall/ CRC, London, 2008.

23. E. Harrison, "A Multiplicative Seasonal Box-Jenkins Model to Nigerian Stock Prices," Interdisciplinary Journal of Research in Business, Vol. 2, No. 4, 2012, pp. 1-7.

24. D. N. Gujarati, "Econometría," McGraw Hill, México, 2003.

CHAPTER 13

Perspectives on Cultural and Sustainable Rural Tourism in a Smart Region: The Case Study of Marmilla in Sardinia (Italy)

Chiara Garau

Department of Civil and Environmental Engineering and Architecture, DICAAR, University of Cagliari, via Santa Croce, 67, Cagliari 09124, Italy

ABSTRACT

This paper is being inserted into the current debate on the topic of sustainability, as it applies to rural tourism. In particular, it addresses the need to identify strategic actions that will enhance the dissemination of cultural resources to facilitate cultural planning. Balancing the dynamic tension that characterizes the relationship between tourism development and protection of the landscape is key to finalizing appropriate planning strategies and actions, especially in the context of marginal rural areas. In support of theoretical and methodological reflections pertinent to this relationship, this paper presents a case study of the region of Marmilla on Italy's island of Sardinia. The absence of both a "cultural planning" philosophy and a strategic approach to systemic and sustainable rural tourism in this country has been acknowledged. This paper concludes by discussing the results that emerged during the preparation of this case study, with respect to smart, sustainable, rural tourism development, while accepting the need for compromises between the force of globalization, nature, tourism, places, and people.

Keywords: cultural tourism; rural tourism; sustainable tourism; smart land use; cultural planning; Sardinia

1. INTRODUCTION

The phenomenon of rural tourism has recently assumed new significance, having risen gradually from a marginal to a widespread practice. Changes to the European countryside, increasing globalization, and growing competition between traditional locations and new destinations have raised tourist expectations, encouraging travel to places less well known, and recalling, for this reason, those tourist flows attracted by the authenticity of the experiences visitors are likely to have [1,2,3,4]. The pursuit of authenticity—consisting of cultural and social identities, traditions, memories, intangible connections, local peculiarities, and rural landscapes—has therefore led communities and local, national, and European governments [5] to respond to new tourism demands in more complex ways [6].

In Europe, this phenomenon has been consolidated over time, beginning in the late eighties. Since then, the literature has documented a wide range of theoretical paradigms aimed at interpreting these consequences, and the natural environment has seen the emergence of rural tourism. One of the first definitions of rural tourism was proposed by the European Commission in 1986: "Rural tourism is a broad concept that includes not only farm tourism or agritourism—accommodation provided by farmers—but all tourist activities in rural areas" [7]. Simonicca (1997) interpreted rural tourism as an alternative type of tourism with sustainable objectives [8]. Namely, socio-cultural or natural rural tourism environments represent alternatives to the places where tourists live, and they in turn try to experience positive and educational impacts from having visited them [9]. Daugstag (2008) defined the discovery of the rural land as "a refuge from urban life" [10]. That is, it represents an alternative to embracing globalization. Barke (2004) argued that rural tourism had developed in response to two factors. These were the "decline of traditional rural activities, principally agriculture and the consequent demographic changes, especially depopulation," and "the perceived need to diversify the [...] tourism product away from traditional mass beach tourism characteristic of the 1960s and 1970s" [11]. During that period, the phenomenon that Henry Lefebre called "the right to the city" [12] was tangible: the population had moved from rural areas to cities, mainly because the city provided the civitas, namely the social ties, functions, and services capable of providing an urban lifestyle. These conditions can occur in areas with small populations.

In the nineties, the population increased, as did the volume of information. Competitive territorial marketing strategies began to appear. They were aimed not only at improving the performance and increasing the attractiveness of the land, but also at maximizing its long-term economic benefits; they were also designed to ensure the sustainability of the territory, through proper management of tourist flows. Especially for rural areas, different forms of place-based enhancements such as eco-museums, farms, community maps (or parish maps), educational farms, and educational tours increased.

Corner and Swarbrooke (2004) emphasized the two main outcomes offered by rural tourism: farm hosting, in which country homes were used to provide hospitality, and farm holidays, during which the tourist was placed in rural areas and participated actively in the rural lifestyle that had been preserved as a primary agricultural activity [13].

Cawley and Gillmor (2008) argued that we could speak of rural tourism when there were strong links with the land's economic and productive activities, and three main features were in evidence: integration, sustainability, and endogeneity [14]. Schubert (2006), on the other hand, believed that rural tourism represented a key strategy for regional development [15]. Zhou (2014) asserted that rural tourism was by nature mainly "domestic, and positioned as a small-scale activity" [16], and Balestrieri (2005) highlighted the versatility of rural tourism, and suggested that it was for this reason that it was able to "play the role of engine of sustainable economic development" [17]. In sum, the rural landscape brought environmental sustainability together with different social, cultural, and economic components, due to its different, content-rich, inter-connected, and integrated attractors, along with its strong anthropological characteristics, because it related to the culture and lifestyles of settled communities, and thereby enabled planning for tourism.

In view of global economic dynamics, planning and programming for the development of sustainable rural tourism represents one of the new challenges for strengthening and revitalizing lands that are otherwise not competitive. This paper begins by defining rural tourism as it has become established in Italy during the last ten years. Then, a case study of the region of Marmilla in Sardinia, Italy is presented. The author elaborates on a process for establishing a sustainable rural tourist destination, and describes the necessary cultural changes required of the different actors and tourist enterprises that would be involved in this process. It is assumed that Marmilla is a model of tourism development, and demonstrates best practices for rural cultural tourism, based on recent "smart region" paradigms [18,19], "Neogeography" [20,21], and the development of computing platforms that are increasingly more integrated and interactive [22]. Finally, sustainable rural tourism is discussed, using the study's findings. The paper concludes by recommending the development of smart rural-urban linkages, and demonstrating how even the planning and programming of rural tourism cannot escape comparisons with sustainability, which should be seen not as a constraint, but as a goal of contemporary management and place-based marketing.

2. SUSTAINABLE RURAL TOURISM IN ITALY: SOME REFERENCE DATA

Although tourism was one of Italy's most dynamic economic sectors in the recession period [23], the tourist presence has fluctuated in recent years. In fact, Italy's tourism sector data have shown a greater loss of market share than those recorded by other direct competitors such as Spain and France [24]. There are various reasons for this decrease: unstructured governance, fragmentary promotions abroad [25], a lack of engagement by the National Agency of Tourism [26], the Italian economic tourism sector being composed predominantly of small and medium-sized businesses [27], the lack of competitive tourism products, poor infrastructure, and a general lack of coordination at the political, technical, and operational decision-making levels [28]. In addition, there are problematic aspects related to the use of digital platforms, such as the lack of a digital tourism strategy, and the insufficient use of digital sales channels and applications for smartphones and tablets. To address these weaknesses, the Strategic Plan for the Development of Tourism in Italy (Piano strategico per lo sviluppo del turismo in Italia (2013)) identifies targeted and specific actions that are to be undertaken: the construction of a laboratory for e-tourism; the development of mobile applications specifically targeted at foreign tourists, who know how to integrate logistical information with cultural resources (such as museums and exhibitions); the definition of roles and responsibilities; and the coordination of governance arrangements, not only between the state and the regions/provinces, but also between provinces and municipalities [29]. Service integration, mobile applications, and the coordination of decision making at all levels are key aspects of national tourism innovations, and they need to be relaunched in a smart way.

These factors have also been of interest in rural areas, where they have produced different results. In these areas, in fact, there have been positive trends, especially from 2003–2013. Although there have been no ad hoc surveys on the phenomenon, and although there are objective difficulties involved in identifying a destination's unique rural character [30], this trend toward a rural tourism sector is apparent from the consistent growth of agritourism farms.

As shown in Table 1, their number rose from 13,019 in 2003, to 20,897 in 2013, thereby increasing by 60.5%. Firms dedicated to other agritourism activities have acquired an important role. They have shown increases of 47.9% and 62.7% for tasting and other activities respectively. In the same period, the beds available have increased by 94,738, and the seating capacity has increased by 157,615 [31]. The major "boom" in 2013, however, concerned activities related to "educational farms", which, compared to 2010—the year they began to appear—have increased by 15%. While central Italy offers agritourism farms with greater diversification (Table 2), in the period between 2012 and 2013, agritourism farms grew more significantly in the north at 6.1%, relative to those in the center and on the islands; in the south, a decline of 5.8% was observed (Table 3) [31]. The regions of Tuscany and Trentino Alto Adige have the most agritourism farms (in 2013 they exceeded 3500 units). Agricultural hospitality is more rooted there for historical reasons. Agritourism farms are also widespread in the regions of Lombardy, Veneto, Umbria, Piedmont, and Emilia Romagna (with over 1000 companies), and, finally, Lazio, Marche, and Sardinia.

Although the data described above tend to underestimate the phenomenon [32], they confirm that rural tourism is a reality with its own identity, and a well-characterized question. Interest in this sector is destined to grow, probably, as mentioned in the previous paragraph, due to evolving tourism preferences and demands. A large part of rural Italian tourism is still very much influenced by several factors, including the fragmentation and lack of cohesion of the valorization choices in an extensive rural area; local communities' disinclination to accept change, being too weak to assume the stresses that accompany innovative tourism; and the fairly limited capacities of existing hospitality structures.

Linking rural tourism with cultural tourism can lead to the integration and diversification of tourism opportunities, and translate the tourist destination as a whole into a competitive destination. In fact, the combination of cultural and rural offerings are able to support political, institutional, or proactive choices for new tourism products, and may represent a unique experience [3,4], not only for the tourist in search of authenticity, but also for all other stakeholders (the local communities, local administrators, institutions, non-profit organizations, and business employees) involved in the territory. Rurality, widespread cultural heritage, and tourism therefore comprise a complex relationship that imposes on territories and local actors the expectation that they will clarify their tourism potential, and the related processes already in place. In this way, the rural area is no longer a "product", but is rather a place where sustainable place-based integrated development processes are activated [33].

Table 1. Typologies of agritourism farms in Italy (ISTAT Data, 2014).

Typologies of agritourism farms	2003	2004	2005	2006	2007	2008	2009	2010	2011	2012	2013	Difference 2003–2013	
												Absolute Number (Abs.)	%
Accommodation													
- Firms	10,767	11,575	12,593	13,854	14,822	15,334	15,681	16,504	16,759	16,906	17,102	6335	58.8
- Number of beds	130,195	140,685	150,856	167,087	179,985	189,613	193,480	206,145	210,747	217,946	224,933	94,738	72.8
- Picnic areas	4540	5386	5826	6935	7055	7320	7785	8759	9113	8363	8180	3640	80.2
Food & beverage													
- Firms	6193	6833	7201	7898	8516	8628	9335	9814	10,033	10,144	10,514	4321	69.8
- Seating capacity	249,342	266,654	277,866	298,003	322,145	337,385	365,943	385,470	385,075	397,175	406,957	157,615	63.2
Tasting													
- Firms	2426	2737	2542	2664	3224	3304	3400	3836	3876	3449	3588	1162	47.9
Food & beverage													
- Firms	6193	6833	7201	7898	8516	8928	9335	9914	10,033	10,144	10,514	4321	69.8
- Seating capacity	249,342	266,654	277,866	298,003	322,145	337,385	365,943	385,470	385,075	397,175	406,957	157,615	63.2
Tasting													
- Firms	2426	2737	2542	2664	3224	3304	3400	3836	3876	3449	3588	1162	47.9
Other Activities													
- Firms	7436	8240	8755	9643	9715	10,354	10,583	11,421	11,785	11,982	12,096	4660	62.7
of which													
- Horse Riding	1364	1494	1478	1557	1559	1615	1548	1638	1662	1489	1230	−134	-9.8
- Excursionism	2452	2692	2981	3131	2879	3140	3071	3190	3233	3324	3124	672	27.4
- Naturalistic Obs.	224	265	575	517	558	697	623	784	891	932	972	748	333.9
- Trekking	1350	1463	1426	1465	1629	1657	1674	1950	1949	1821	1717	367	27.2
- Mountain Bike	2101	2422	2258	2311	2347	2398	2309	2800	2794	2785	2851	750	35.7
- Educational Farms	-	-	-	-	-	-	-	752	1122	1251	1176	1176	-
- Courses	693	812	942	1025	1256	1407	974	1967	1878	2009	1770	1077	155.4
- Sports	2927	3806	3474	3682	3758	4203	4168	4152	4141	5058	5088	2161	73.8
- Various	3786	4003	4288	5043	5395	5616	5994	6312	6737	4917	6033	2247	59.4
Agritourism Farms													
- Total Firms	13,019	14,017	15,327	16,765	17,720	18,480	19,019	19,973	20,413	20,474	20,897	7878	60.5

Sardinia was chosen for this study for several reasons. First, this island's rural tourism processes have never before been considered as offering an alternative, more consolidated development model relative to the seaside model. Rural tourism has always been regarded as secondary to seaside tourism. The development of a rural development model, however, could facilitate the attenuation of Sardinia's imbalances by countering the seasonal ebbs and flows characteristic of seaside tourism. Secondly, Sardinia ranks eleventh in the total number of Italian farms by region, and represents the median number of farms (Table 3). For these reasons, Sardinia is considered to be representative of the rural Italian reality. This case study of the natural region of Marmilla in Sardinia also addresses questions associated with integrating the cultural, natural, and rural heritage. Even though the Marmilla region has this significant potential, a strategic process for tourism planning has not yet been created. In this paper, based on the following concepts: tourism, rurality, and cultural heritage, the author defines the conditions needed to initiate a process of reflection that could lead to defining a strategic tourism planning process.

Table 2. Agritourism farms in Italy by type of activity and region (ISTAT Data, 2014).

Regions Geographical distributions	Food & beverages				Tasting				Other Activities			
	2012	2013	Difference Abs.	%	2012	2013	Difference Abs.	%	2012	2013	Difference Abs.	%
Piedmont	753	790	37	4.9	589	616	27	4.6	902	925	23	2.5
Aosta Valley	45	36	−9	−20.0	35	9	−26	−74.3	10	9	−1	−10.0
Lombardy	1019	1060	41	4.0	116	144	28	24.1	673	722	49	7.3
Trentino-Alto Adige	577	625	48	8.3	100	108	8	8.0	1311	1348	37	2.8
Bolzano-Bozen	430	470	40	9.3	-	-	-	-	1255	1292	37	2.9
Trento	147	155	8	5.4	100	108	8	8.0	56	56	-	-
Veneto	756	782	26	3.4	603	641	40	6.7	511	524	13	2.5
Friuli-Venezia Giulia	447	454	7	1.6	10	13	3	30.0	229	240	11	4.8
Liguria	281	353	72	25.6	-	40	40	-	336	287	−49	−14.6
Emilia-Romagna	797	834	37	4.6	-	-	-	-	874	739	−135	−15.4
Tuscany	1131	1232	101	8.9	577	515	−62	−10.7	2925	3141	216	7.4
Umbria	405	409	4	1.0	227	237	10	4.4	1108	1120	12	1.1
Marche	414	447	33	8.0	380	420	40	10.5	306	234	−72	−23.5
Lazio	551	596	45	8.2	134	163	29	21.8	552	571	19	3.4
Abruzzo	436	410	−26	−6.0	73	56	−17	−23.3	467	377	−90	−19.3
Molise	86	86	-	-	50	50	-	-	54	54	-	-
Campania	352	396	44	12.5	136	151	15	11.0	287	330	43	15.0
Puglia	271	222	−49	−18.1	146	138	−8	−5.5	231	303	72	31.2
Basilicata	98	78	−20	−20.4	40	20	−20	−50.0	104	54	−50	−48.1
Calabria	565	542	−23	−4.7	50	49	−1	−2.0	503	472	−11	−6.2
Sicily	473	493	20	4.2	186	219	33	17.7	514	550	36	7.0
Sardinia	683	669	−14	−2.6	-	-	-	-	85	96	11	12.9
ITALY	10,144	10,514	370	3.6	3449	3588	139	4.0	11,982	12,096	114	1.0
Northern Italy	4675	4934	307	6.6	1451	1571	128	8.8	4846	4794	−15	−0.3
Central Italy	2501	2684	183	7.3	1317	1334	17	1.3	4891	5066	175	3.6
Mezzogiorno	2968	2896	−27	−8.9	681	683	31	4.6	2245	2236	10	0.4
South Italy	1812	1734	−78	−4.3	495	464	−31	−6.3	1646	1590	−56	−3.4
Islands	1156	1162	6	0.5	186	219	33	17.7	599	646	47	7.8

Table 3. Agritourism farms in Italy by region (ISTAT Data, 2014).

Regions Geographical distributions	2012	2013	Difference Abs.	%	Regions Geographical distributions	2012	2013	Difference Abs.	%
Piedmont	1164	1220	56	4.8	Umbria	1262	1280	18	1.4
Aosta Valley	54	53	−1	−1.9	Marche	788	880	92	11.7
Lombardy	1415	1521	106	7.5	Lazio	841	884	43	5.1
Trentino-Alto Adige	3391	3506	115	3.4	Abruzzo	774	653	−121	−15.6
Bolzano-Bozen	2996	3098	102	3.4	Molise	104	104	-	-
Trento	395	408	13	3.3	Campania	407	458	51	12.5
Veneto	1376	1449	73	5.3	Puglia	355	353	−2	−0.6
Friuli-Venezia Giulia	588	614	26	4.4	Basilicata	145	112	−33	−22.8
Liguria	543	567	24	4.4	Calabria	610	577	−33	−5.4
Emilia-Romagna	1036	1106	70	6.8	Sicily	602	633	31	5.1
Tuscany	4185	4108	−77	−1.8	Sardinia	834	819	−15	−1.8
ITALY	**20,474**	**20,897**	**423**	**2.1**	**Mezzogiorno**	**3831**	**3709**	**−79**	**−2.1**
Northern Italy	**9567**	**10,036**	**534**	**6.1**	**South Italy**	**2395**	**2257**	**−138**	**−5.8**
Central Italy	**7076**	**7152**	**76**	**3.1**	**Islands**	**1436**	**1452**	**16**	**1.1**

3. GOVERNANCE AND MANAGEMENT OF MARMILLA'S PLACE-BASED HERITAGE FROM A SUSTAINABILITY PERSPECTIVE

Based on the preceding information, the question that arises is how to initiate a sustainable tourism approach for an agricultural destination that does not yet have a strategic planning and management approach for its tourism sector. We begin this research by analyzing Marmilla's rural and natural landscape, as well as its cultural and historic heritage. Doing so allows us to understand the extent and value of Marmilla's cultural and environmental aspects. Next, socio-demographic and production supply and demand dynamics are analyzed, to help delineate an appropriate strategic policy for place-based development and tourism planning. Finally, using different scales, the application of direct and indirect governance instruments as they might affect the development of tourism and environmental protection in the Marmilla context are analyzed. These analyses reveal that a successful strategic planning and management approach for Marmilla depends primarily on social capital. Indeed, awareness of local administrators regarding their roles and the potential of a strategic approach to systemic and sustainable rural tourism will provide the conditions needed to improve the attractiveness (and therefore also the place-branding) of the cultural and natural heritage, and make a significant contribution in terms of political cohesion, identity and local development. In this regard, it has been observed that the role of technology can be crucial to the debate on governance for sustainable tourism in rural areas. Indeed, Go, Della Lucia, Trunfio, and Martini (2014) state that there is the need for a link between "two broad knowledge domains, ICTs and place branding [of rural contexts], often isolated from one another" [34]. Tying these two aspects does not delegate to rural areas the characteristic of "sustainable place", but offers them the opportunity to have a voice, visibility and place-based development in the era of globalization [34]. In addition, it captures the need to overcome certain structural limitations, creating, for example, a "system of rural areas": especially in Marmilla, a recent study showed that four of its 18 municipalities (Genuri, Tuili Turri, and Ussarramanna) are at risk of demographic desertification [35].

For this reason, in this paper, with the help of new technologies, the most relevant factors to emerge from previous analyses are identified, collected, and "filtered", using a logical framework, and hypothesizing a scenario aimed at two different but parallel objectives: protecting the natural and cultural heritage, and advancing sustainable tourism development in the case study region of Marmilla, while accepting the need for compromises between the forces of globalization, nature, tourism, places, and people.

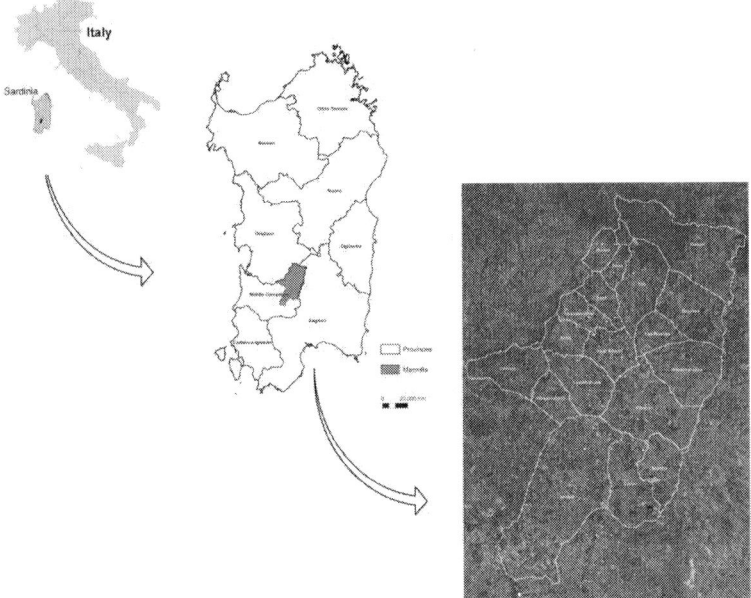

Figure 1. The natural region of Marmilla.

Figure 2. Marmilla's demographic trend from 1861 to 2011 [41].

3.1. Case Study of Marmilla

The region of Marmilla is located in the southern-central part of Sardinia, Italy, in the province of the Middle Campidano (Figure 1). It covers 415 square kilometers, and had a population of 25,619 in 2014 [36].

Marmilla's morphology is mainly devoted to agriculture, supported by a system of small settlements. It comprises 18 municipalities [37] and, as shown in Table 4, is characterized by features typical of southern Italy's inland areas: a low population density, weak economic structures, negative long term demographic trends (Figure 2 and Table 5), and an aging population [38,39,40]. The social and demographic situation is not uniform, and some of the 18 municipalities do not exceed 300 inhabitants (such as Setzu and Las Plassas with 151 and 240 inhabitants respectively), while others have more than 8500 inhabitants (Sanluri). The territory is characterized by a strong reliance on commuting, especially closer to major urban centers such as Cagliari, Oristano, and Sassari.

Table 4. Socio-demographic and economic dynamics in Marmilla [42].

Geographical distributions	Area (km square)	Population density (2014)	Population 2014 (inhabitants)	Aging index 2014	Disposable Income Per capita 2013 (Income—Taxes [fiscal levy])
Barumini	26.57	48.78	1296	261.4	€ 11,986.00
Collinas	20.79	41.41	861	292.9	€ 12,507.00
Furtei	26.12	64.20	1677	204.8	€ 11,117.00
Genuri	7.55	45.03	340	463.6	€ 12,855.00
Gesturi	46.87	27.29	1279	319.4	€ 10,596.00
Las Plassas	11.14	22.08	246	255.6	€ 10,596.00
Lunamatrona	20.57	85.03	1749	277.7	€ 13,028.00
Pauli Arbarei	15.12	42.72	646	247.7	€ 11,465.00
Sanluri	83.78	101.81	8530	174.5	€ 14,418.00
Segariu	16.69	74.18	1238	206.9	€ 10,596.00
Setzu	7.82	19.31	151	273.7	€ 14,071.00
Siddi	11.02	61.43	677	321.2	€ 11,291.00
Tuili	24.5	42.86	1050	375.5	€ 11,986.00
Turri	9.64	46.37	447	442.1	€ 11,986.00
Ussaramanna	9.75	57.23	558	305.0	€ 13,376.00
Villamar	38.64	72.93	2818	174.0	€ 11,465.00
Villanovaforru	10.97	59.16	649	272.3	€ 11,986.00
Villanovafranca	27.46	51.24	1407	283.1	€ 10,944.00
Marmilla	415	61.73	25,619	226.5	€ 12,014.94
Sardinia	24,090	69.07	1,663,859	174.4	€ 13,871.00
Italy	301,338	201.71	60,782,668	154.1	€ 17,038.20

Table 5. Marmilla's demographic trend from 1861 to 2011 [41].

Geographical distributions	1861	1871	1881	1901	1911	1921	1931	1936	1951	1961	1971	1981	1991	2001	2011
Baronana	1314	1187	1221	1118	1179	1335	1445	1431	1685	1729	1647	1516	1423	1413	1310
Collinas	976	1013	1072	1033	1088	1065	1046	1091	1206	1213	1129	1145	1076	1014	885
Furtei	1030	915	981	1057	1138	1179	1280	1412	1728	1846	1788	1830	1729	1723	1674
Gesturi	359	400	434	383	440	446	535	575	654	706	567	518	444	386	345
Gestun	1669	1457	1430	1431	1507	1455	1643	1709	1827	1801	1567	1515	1438	1430	1280
Las Plassas	485	459	439	397	444	500	547	502	566	632	379	298	391	269	257
Lunamatrona	968	1018	1104	1148	1389	1278	1467	1640	1948	2017	1850	1886	1895	1858	1785
Pauli Arbarei	433	434	469	401	477	530	556	676	801	797	787	778	692	730	651
Samhut	4199	4177	4177	4403	4593	4786	5449	5721	7555	7595	7402	8305	7912	8519	8460
Segariu	700	588	647	661	732	750	899	989	1308	1441	1409	1432	1320	1358	1277
Setzu	303	330	276	340	292	267	304	325	278	278	233	223	184	166	144
Siddi	608	615	605	636	631	802	869	871	987	1121	990	903	869	799	696
Tush	1334	1286	1242	1320	1330	1302	1478	1613	1713	1591	1348	1347	1283	1185	1082
Tursi	454	488	511	503	490	487	575	602	729	734	633	597	572	533	442
Ussaramanna	623	540	509	586	677	730	790	863	920	963	835	714	656	611	556
Villamar	1948	1825	1903	2047	2250	3220	2675	2876	3301	3389	3657	3196	3147	3060	2872
Villanovaforru	506	517	593	615	635	709	741	770	905	931	846	789	733	700	681
Villanovafranca	1356	1166	1121	1139	1286	1369	1577	1633	2055	2117	1759	1871	1621	1491	1433
Marmilla	19,156	18,504	18,762	19,168	20,498	21,210	24,810	26,309	30,166	30,881	28,316	28,873	27,166	27,136	25,808
Sardinia	609,015	636,413	680,460	795,793	868,181	885,467	983,700	1,034,206	1,276,023	1,419,362	1,473,800	1,594,175	1,648,248	1,631,880	1,639,362

Extensive territories with high environmental quality and low population levels contribute to conserving the land over time, and maintaining its rurality (Figure 3). On the other hand, the region lacks a socially young and dynamic cohort. These issues have become structural problems for place-based development in Marmilla, and have led the Marmilla region to be perceived as being on the margins of Sardinia's economy.

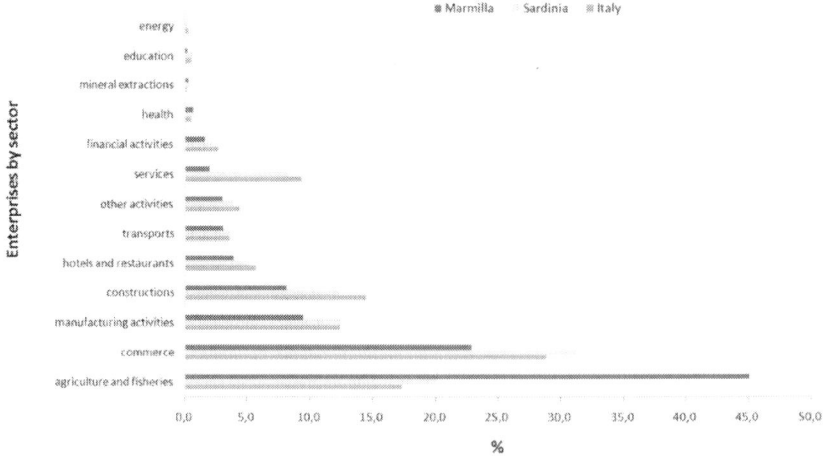

Figure 3. Segmentation by percentage of enterprises (year 2010) [43].

In this context, the development of tourism could limit the impacts of the critical weaknesses described above, by offering new land-based job opportunities to members of the local community, thereby benefitting the entire economy and society. Natural, cultural, and environmental attractions abound:

the uniqueness of the landscape, the historical value of the small urban centers, and the wealth of traditions linked to the agricultural economy, united by the presence of many archaeological and nuraghic remains. Among the latter, the Nuraghe of Barumini (included in the United Nations Educational, Scientific and Cultural Organization's [UNESCO] list of World Heritage Sites in 1997) is a prime destination in the region of Marmilla, and its draw benefits the whole island of Sardinia. However, tourist groups that visit the Nuraghe often do not consider the island's many other attractions, and so the Nuraghe do not have an appreciable economic impact on the extended area, in terms of employment or stimulating commercial activities.

The municipalities in Marmilla have tried in recent decades to adopt a systematic approach to building a local shared identity, by establishing numerous associations [44]. Government administrators have tried to diversify tourism sites by proposing—unfortunately only on paper—the integration of natural attractions (such as the Geobotanical Park, the Museum of the Territory, and the Jars of Siddi, Collinas, Tuili, and Setzu), with archaeological sites (such as the various archaeological museums and the many areas in which Nuraghes, sacred wells, and the Tombs of Giants are present), and with folkloristic, ethnographic, demo-ethno-anthropological, and historical points of interest (small museums, Romanesque churches, altarpieces from the sixteenth century, cultural centers, festivals and village traditions) [45].

In other words, local administrators have not managed to effectively organize a real "team" that could assume responsibility for the formulation of a strategic plan for place-based tourism, using place-based logic to enrich the local community's purely parochial visions. Maybe this is because (temporary) local ambitions promoted place-based identities that did not represent the local community in the best possible light: the area of Marmilla, as mentioned before, has historically been dedicated to agriculture and pastoralism. Not including these aspects in the range of tourist offerings has definitely compromised its place-based, economic, and social development. It is clear that the context for development will not be mass tourism, but rather it will have to consider the interrelationships between sustainability, the local culture, the rural environment, the community, and tourism. The peculiarities of the rural land must remain unchanged, without becoming part of a vicious cycle in which they are sacrificed for economic development.

Other Italian regions (such as Umbria and Marche) have managed to create place-based brands for some areas, and have therefore been able to offer competitive tourism that is linked to the natural characteristics of the agricultural and mountainous sectors.

The typical quality of life in rural areas, the reasonable cost of living, and its central location in Sardinia could combine to make Marmilla a strategic tourist destination, if positive hospitality, commercial, and tourism sector processes are actually implemented.

3.2. Development of the Tourism Sector in Marmilla

According to published literature, sustainability has environmental, social, and economic dimensions [46,47]. Tourism development is thus sustainable only if has been strategically planned to achieve goals that manifest their effects over the long term. For this reason, Marmilla's strategic programming framework was analyzed at different scales, considering various government instruments such as the Rural Development Programme (RDP) 2014–2020 [48], the Regional Development Plan (RDP) 2014–2019 [49], urban master plans (for all eighteen municipalities), and the spatial strategic plan of Sanluri (the only municipality in Marmilla using this urban tool).

This comparative analysis facilitated the development of a vision for strategic and planned processes in the case study area. It also contributed to an understanding of how best to enhance integrated planning policies for a public/private initiative, and, most importantly, how to requalify rural lands for landscape protection, and for cultural and sustainable rural development (Table 6).

Table 6. Evaluation criteria for redeveloping the area of Marmilla.

Protection and Cultural Landscape	Development of Sustainable Tourism
ENVIRONMENTAL SUSTAINABILITY; Preservation of landscapes, habitats, and ecosystems; Promotion of the use of renewable resources; Introduction or improvement of environmental management systems; Protection of the main territorial vocations	LOCAL ECONOMIC DEVELOPMENT; Encouraging the development of local firms and businesses; Encouraging the formation of employment aimed at sustainability of the tourism sector; Encouraging public and private partnerships; Promoting the construction or renovation of buildings for rural tourism, sustainable in the long term, despite changing political mandates
ECONOMIC SUSTAINABILITY; Development of landscape quality recognized by international bodies (UNESCO, etc.); Development of a market for local goods and sustainable services; Encouraging investments in innovative, environmentally friendly technologies	SUSTAINABLE ENVIRONMENTAL DEVELOPMENT; Encouraging initiatives dedicated to diversification of tourism and to the redistribution of tourist flows; Protecting and promoting the cultural and historical heritage; Encouraging the demand for and the achievement of environmental certification in the tourism sector
SOCIAL SUSTAINABILITY; Triggering processes of awareness of local tangible and intangible goods; Activating processes to raise awareness of the topic of sustainable development, protection, and enhancement of cultural heritage and public spaces; Improving the participatory practices; Improving the network of relationships between various stakeholders	SOCIAL DEVELOPMENT; Developing the integration between different policy sectors; Construction of participatory practices aimed at promoting the latest information technology tools for tourism purposes; Developing a place-based approach in which the local community is integrated with tourists; Promoting opportunities that induce residents to identify unique regional elements

The principles summarized in Table 6 provide insight into how best to orient local communities in Marmilla to these processes. However, the most important innovations for triggering development in this rural area are of two types. These are the region's social capital, and the latest technologies capabilities for raising awareness of the area's rural realities worldwide, and making them attractive for tourists interested in this niche sector.

The local community plays a crucial role in providing sustainable social capital, and must therefore participate in the development process for the long term [50]. Entrepreneurs will be able to increase the area's place-based competitiveness with their innovations [51]. Go, Della Lucia, Trunfio, and Martini (2014) emphasize how important it is to consider social capital under the structural [52], cognitive [53], and relational [54] dimensional perspectives.

For these reasons, particularly in Marmilla there is a need to maintain the area's traditional and historical vocational aspects [55]. Tourists' destination choices in turn may lead to a change of original vocation and increasing competitiveness—but only temporarily (in relation to the structural perspective). It is also important that Marmilla's administrators, including legislators, coordinators, and promoters, start the development process, while conducting their different roles and responsibilities, in line with the dimensional perspective [56]. And finally, the local population's "internal" visions and tourists' "external" visions for a rural place—that constitute the relational dimension of social capital—can help to define a common agenda. In so doing, they will help to identify goals and objectives for developing and promoting the area. For that reason, it is also important to use today's communication and marketing technologies to promote and embed knowledge of the attractiveness of the local area in national and international tourist markets. With regard to the latter point, it will be helpful to identify incentives associated with particular ecological brands recognized by the World Trade Organization (WTO) in the international context such as Ecolabel, or Orange Flag; Organic Farms; Associazione Borghi Autentici d'Italia (The Association of Italian Authentic Villages); and I Borghi più Belli d'Italia (The most Beautiful Villages of Italy), promoted by the National Association of Italian Municipalities (ANCI) in the national context.

There have been some sporadic adhesions of this type in Marmilla's municipalities. Collinas joined The Association of Italian Authentic Villages. The farm su Massaiu of Turri, which represents an example of excellence at the local level, has joined Ecolabel. In fact, su Massaiu produces a range of local melons, saffron, and a type of grain called Senatore Cappelli. Orchards, olive groves, almonds, medicinal herbs, legumes, and vegetables in general are also cultivated. The farms serve these products to customers or sell them. The benefits of owning such brands are significant. External communications are strengthened by combining the brand's organization (or brand's destination) with the brand's certification, and a virtual path is taken to manage the activities necessary for achieving and maintaining such recognition.

Regarding the second point, using the latest ubiquitous technology to foster, external visibility can do much to leverage development of the Marmilla area, and open the way for innovation, without sacrificing the local cultural and environmental identity. This hypothesis, currently in the concept stage, is further discussed in the following section.

4. TOURISM AND NEW MEDIA, A POSSIBLE COMBINATION IN RURAL CONTEXTS?

In recent years, global organizations—including the International Council on Monuments and Sites (ICOMOS), UNESCO, and the WTO—have promoted cultural tourism [57], using new technologies [58]. Innovative economic development projects aimed at creating "Smart Regions" [18,19], the application of tools associated with "Neogeography" [20,21], and the development of more integrated and interactive computing platforms [22], have led to many experiments, including some in the field of tourism. Over the last decade, the main innovations in this field have included the creation of specific Internet portals and smart cards, fostering a locale's cultural heritage, and, finally, the diffusion of mobile tourism applications. Technology has allowed the formation of more dynamic and "immersive" relationships between tourists, the area's cultural heritage, and the region [58,59].

Ever-present technology has impacted consumers' demands for vacations, as well as the travel and tour packages being offered by suppliers. It has also enabled the diffusion of tourism information in a manner that has ensured maximum customization and user interaction, in contrast to the influence of traditional paper maps or guidebooks on these activities [59].

With continuing innovations, the information communication technologies (ICT), social networks, and mobile applications can facilitate the integration of tourism products in different moments of the consumption process. They have opened the doors to online marketing, strategic management, and marketing information systems (MkIS). They have changed how tourists search for information on different places and evaluate their alternatives. Service providers can now conduct surveys on tourism behaviors before and after the tourists select a desired destination.

However, in a rural and marginal area such as Marmilla, which is firmly anchored in tradition, the use of technological tools will have to be well understood and supported by the local community if they are to be accepted. These tools appear to be effective when they are accompanied by continuous training for the local communities, thereby creating a self-renewing and reproducing process over time.

In Marmilla then, cultural, rural, and natural excellence needs to be identified, communicated (through various tools, from online to offline dashboards), and systematized, and then used as part of a strategic approach for tourism governance. Porrello (2006) also included cultural planning as an essential part of the overall perspective [60]. In Italy, there are sporadic cases in which applications downloadable on smartphones or tablets have been used to promote tourism in rural areas. Of Italian regions, Tuscany has been the most active [61], and even if their communications through digital apps are not organized in an interactive way, they look very similar to website consultations. This mode of technological fruition in rural places is still not widespread in Sardinia. For this reason, the organized and attractive simulation of rural tourism paths that have been realized using the Garau's model [19,58,59] is proposed.

The implementation of rural pathways has led to thinking of whole, place-based involvement, as part of a strategic logic network that will enhance the resources of individual municipalities for the benefit of the whole region of Marmilla. On the basis of the analyses conducted here (Section 3.1), and considering the need to improve social and technological capital (Section 3.2), the author has identified a number of preliminary strategic actions that are linked to the enhancement of the area's cultural and rural heritage, and to the development of sustainable tourism. These actions can be considered as a set of best practices, and a place for local administrators to start, before beginning work on a cultural planning tourism development model.

The recommended actions are listed below:

1. There is an overarching need to promote sustainable tourism across Marmilla, and not only in the main polarity (for instance what today includes the Nuraghe of Barumini), by highlighting the region's rural and cultural aspects and creating an integrated quality. These aspects include local museums, exhibitions, cultural events, and the interrelation of individual museums associated with rural structures, and interactive educational farms and/or guided tours as part of ongoing rural processes—such as harvesting grain and saffron.

2. A second required action is developing the skills needed to begin proper planning and programming for rural tourism, through a centralized control, in which experts do not have local interests and are capable of supporting the often fragmentary and conflicting dynamics typical of rural areas. This can be achieved by involving experts who are external to the region.

3. Next, it will be necessary to enhance local places' competitiveness (for example, by returning to traditional ways of promoting eco-museums, permanent, temporary, or itinerant events, exhibitions, and installations) and entrepreneurial tourism (through prizes or incentives to operators and companies that are distinguished by the quality offered, or through actions that stimulate an increase in the number of beds offered, while basing projects on ethical and sustainable development models).

4. A fourth action will involve integrating agricultural activities with services that are compatible with tourism activities, by proposing shop windows that display local products.

5.   The visibility of local attractiveness should be unified, by creating a place-based promotion of the entire region of Marmilla (from the establishment of a centralized structure that encourages the formation of networks of rural enterprises, to joint agreement on a unique logo for the area and/or for place-based marketing).

6.   Demand loyalty should be strengthened, through a series of actions aimed at enhancing internal communications among municipalities, and external communications between Marmilla and the rest of the world (for example, launching marketing actions on specific segments; promoting new tourism packages; conducting surveys to understand visitors' motivations for coming, enhancing the interest and attractiveness of new offers; improving web marketing actions; promoting seasonal offers; and encouraging the movement of visitors from established attractions to previously unknown places of interest).

7.   Finally, a commitment to sustainability should be used as the parameter for planning tourism in the area. On this point, it is essential that common goals and partnerships among the parties involved be identified, to foster understanding and to periodically update the processes, strategies, and planning associated with tourism development.

These actions, which are summarized in Figure 4, can be merged with the definition of rural pathways for Marmilla's place-based tourism redevelopment. Furthermore, the author has identified ways to enhance the quality and integration of local resources, while on the one hand respecting their historical identity, and on the other hand trying to better understand cultural offerings appropriate for rural tourism. This was achieved by reviewing existing literature on the area's cultural heritage and rural offerings [62,63,64,65,66], and from informal discussions with residents and administrators.

The notion of using different pathways, currently in the concept stage, touches the main points of interest (POIs) with a series of customizable stops that are linked to the availability of opportunities to taste local products, visit the historical, architectural, and archaeological museums representative of Marmilla's widespread heritage, and to visit landscaping, natural, and environmental landmarks and permanent agritourism facilities or educational farms across the region.

The proposal for the pathways is premised on the development of an online platform, downloadable from a smartphone or tablet, but supported by printed materials, and capable of being manipulated by the local community. Media dissemination of mobile technologies could attract people who would not otherwise be aware of Marmilla's potential.

The proposed paths would thus create smart rural-urban linkages, and demonstrate how the planning of and programming for rural tourism cannot escape a comparison with sustainability. Furthermore, in a strategic planning process for the tourism sector, the development of more attractive local organizations could lead to advantages for the whole island, by creating tourism districts for example, and increasing, consequently, the attractiveness of the regional offer [67].

Figure 4. Rural pathways in Marmilla.

5. CONCLUSIONS

The challenge that has become apparent by analyzing the Marmilla case study is the need to pursue the goals of place-based improvements, competitiveness, and sustainability in the longer term for this land-locked and marginal area of Sardinia. This perspective—extensible to other national or international rural contexts—is aimed at increasing the number of rural tourist destinations available today, considering the context of Marmilla as a smart tourist destination, and accepting necessary "compromises" between globalization, nature, tourism, places, and people.

As has been discussed, tourism has been used to promote the development of the region of Marmilla by promoting its strengths (the Nuraghe of Barumini), and making improvements that, until now, have been linked to less "traditional" tourist flows. In fact, the Nuraghe of Barumini is a well-known site in the tourism market, but that site alone is unable to trigger new endogenous development across Sardinia.

The integration of the various factors described in this work—such as rural lands, the cultural, archaeological, and natural heritage, local and traditional food production, and networks between all the actors—can constitute a winning tourism offering for tackling the socio-economic issues of rural areas in general, and of the agricultural sector in particular. Tourism structured in this way can trigger collaborative processes among Marmilla's different administrative offices, and thereby initiate actions that will precipitate endogenous development. Namely, they can create the conditions necessary for future developments that include "aggregating tourist offerings", by integrating or diversifying their offerings on the basis of new and different targets such as school tourism, tourism of the third age, religious tourism, and folkloristic tourism.

Strong values consistent with environmental sustainability, and proper diversification of the social fabric and local business community are keys to achieving an essential balance in the case study presented. This vital process must be based on a centralized control that can interface with all the actors, promote their potentialities, and mediate any conflicts. Fragmentary and parochial visions typical of rural areas inevitably lead to heterogeneous approaches to planning and programming tourism products [68]. This difficulty has to be faced and overcome in order to develop tourism products in a time of globalization, when tourists can experience all offerings for a single destination as unique, and cannot be focused on one firm or a single cultural or archaeological product. The image of a destination that emphasizes its vocations and its local cultural heritage is competitive and attractive in itself. It is important to reach the right target, by focusing less on marketing individual products, and emphasizing the uniqueness of the emotional experiences offered.

New forms of communication such as the Internet, social networks, and applications downloadable on smartphones or tablets, such as the one presented in this paper, are essential to improving the visibility and enhancing the promotion of tourism opportunities available in rural Marmilla. However, in this area, where some aspects of a digital divide remain, this form of communication has to be accompanied by formal and informal offline networks. Both modalities of communication are needed to promote the area in the long term, without relying on occasional advertising campaigns. Nevertheless, we must not forget that the successful integration of rural and cultural tourism can be achieved if there is agreement among the various stakeholders—including the local and regional administrators and the local community—to coordinate and encourage the development of local resources. They can, in fact, intervene and help the local tourism sector of Marmilla evolve for the better.

REFERENCES AND NOTES

1. McCannell, D. The Tourist: A New Theory of the Leisure Class; University of California Press: Oakland, CA, USA, 1976.
2. Ferrari, C.; Adamo, G.E. Autenticità e risorse locali come attrattive turistiche: Il caso della Calabria. Sinergie 2012, 66, 79–113.

3. Stamboulis, Y.; Skayannis, P. Innovation strategies and technology for experience-based tourism. Tour. Manag. 2003, 24, 35–43.

4. Turco, A. Turismo e Territorialità. Modelli di analisi, strategie comunicative, politiche pubbliche; Unicopli: Milano, Italy, 2012. (In Italian)

5. Hall, D.R.; Roberts, L.; Mitchell, M. New Directions in Rural Tourism; Ashgate: Hants, UK, 2005.

6. Long, P.; Lane, B. Rural tourism development. In Trends in Outdoor Recreation, Leisure and Tourism; Gartner, W.C., Lime, D.W., Eds.; CAB International: Wallingford, UK, 2000; pp. 299–308.

7. European Commission. Community action in the field of tourism. Commission communication to the Council. Bulletin of the European Communities, Supplement 4/86, p. 10. Available online: http://aei.pitt.edu/5410/ (accessed on 21 February 2015).

8. This concept resumes the sustainable development process, in which current needs are satisfied without compromising the needs of future generations (Brundtland Commission. World Commission on Environment and Development Sustainability; Oxford University Press: Oxford, UK, 1987).

9. Simonicca, A. Antropologia del turismo, strategie di ricerca e contesti etnografici; La Nuova Italia Scientifica: Roma, Italy, 1997; pp. 169–173.

10. Daugstad, K. Negotiating landscape in rural tourism. Ann. Tourism Res. 2008, 35, 402–426.

11. Barke, M. Rural tourism in Spain. Int. J. Tour. Res. 2004, 6, 137–149.

12. Lefebvre, H. The right to the city. In Writings on Cities; Kofman, E., Lebas, E., Eds.; Blackwell: Cambridge, MA, USA, 1995; pp. 63–184.

13. Corner, S.; Swarbrooke, J. International Cases in Tourism Management; Butterworth-Heinemann: Oxford, UK, 2004.

14. Cawley, M.; Gillmor, A.D. Integrated rural tourism: Concept and practice. Ann. Tour. Res. 2008, 35, 316–337.

15. Schubert, D. Active regions–shaping rural futures: A model for new rural development in Germany. In Coherence of Agricultural and Rural Development Policies; Diakosavvas, D., Ed.; OECD Publishing: Paris, France, 2006; pp. 333–352.

16. Zhou, L. Online rural destination images: Tourism and rurality. J. Destin. Mark. Manag. 2014, 3, 227–240.

17. Balestrieri, A.D. Il turismo rurale nello sviluppo territoriale integrato della Toscana; Irpet: Firenze, Italy, 2005.

18. Giffinger, R.; Fertner, C.; Kramar, H.; Kalasek, R.; Pichler-Milanović, N.; Meijers, E. Smart Cities: Ranking of European Medium-Sized Cities; Vienna University of Technology: Vienna, Austria, 2007.

19. Garau, C. Smart paths for advanced management of cultural heritage. Reg. Stud. Reg. Sci. 2014, 1, 286–293.

20. Turner, A. Introduction to Neogeography; O'Reilly: Sebastopol, CA, USA, 2006.

21. Goodchild, M.F. Twenty years of progress: GIScience in 2010. J. Spat. Inf. Sci. 2010, 1, 3–20.

22. Su, K.; Li, J.; Fu, H. Smart city and the applications. In Proceedings of 2011 International Conference on Electronics, Communications and Control, Zhejiang, China, 9–11 September 2011; pp. 1028–1031.

23. Basile, F. Una breve analisi congiunturale del Mezzogiorno nel contesto internazionale e riflessioni preliminari sul settore turistico. In Turismo e territorio: l'impatto economico e territoriale del turismo in Campania; Bencardino, F., Ed.; Franco Angeli: Milano, Italy, 2010; pp. 15–36.

24. Piano strategico per lo sviluppo del turismo in Italia. p. 10. Available online: http://www.agenziademanio.it/export/ download/ demanio/agenzia/5_Piano_strategico_ del_Turismo_2020.pdf (accessed on 21 February 2015).

25. Angeloni, S. Cultural tourism and well-being of the local population in Italy. Theor. Empir. Res. Urban Manag. 2013, 8, 17–31.

26. Musarò, P. Responsible tourism as an agent of sustainable and socially-conscious development: Reflections from the Italian case. Recerca: revista de pensament i anàlisi 2014, 15, 93–107.

27. Sciarelli, S.; Della Corte, V. Il comportamento del turista in condizioni di forte incertezza decisionale. Sinergie rivista di studi e ricerche 2011, 66, 137–152.

28. Paniccia, P.; Vannini, I. Da impresa agricola a agriturismo: un percorso nell'ottica della sostenibilità. In Le imprese nel rilancio competitivo del Made e Service in Italy: settori a confront; Ciappei, C., Padroni, G., Eds.; Franco Angeli: Milano, Italy, 2012; pp. 59–71.

29. Piano strategico per lo sviluppo del turismo in Italia. pp. 5, 30, 47. Available online: http://www.agenziademanio. it/export/download/demanio/agenzia/5_Piano_strategico_del_Turismo_ 2020.pdf (accessed on 21 February 2015).

30. Santamato, V.R. Turismo e territorio: un rapporto complesso. In Esperienze e casi di turismo sostenibile; Messina, S., Santamato, V.R., Eds.; Franco Angeli: Milano, Italy, 2012.

31.   Istat. Le aziende agrituristiche in Italia 2013. Rapporto aggiornato al 10 ottobre 2014. Available online: http://www.istat.it/it/archivio/133966 (accessed on 21 February 2015).

32.   The ISTAT surveys consider only those farms that are treated as such, according to different regional standards. This means that these surveys do not consider those structures that provide the same services but are not incorporated because they are not considered as such by the regional laws.

33.   Papotti, D. Marketing territoriale e marketing turistico per la promozione dell'immagine dei luoghi. Rivista Geografica Italiana 2006, 113, 285–306.

34.   Go, F.; Della Lucia, M.; Trunfio, M.; Martini, U. Governing sustainable tourism: European networked rural villages. In Rural Cooperation in Europe: In Search of the "Relational Rurals"; Kasabov, E., Ed.; Palgrave Macmillan: Chennai, India, 2014; p. 163.

35.   Puggioni, G.; Atzeni, F. I comuni sardi a rischio di estinzione. In Comuni in estinzione. Gli scenari dello spopolamento in Sardegna; Progetto IDMS—2013; Regione Autonoma Della Sardegna: Cagliari, Italy, 2013; pp. 15–45.

36.   Elaborations made by the author on Demoistat Data (2014).

37.   Barumini, Collinas, Furtei, Genuri, Gesturi, Las Plassas, Lunamatrona, Pauli Arbarei, Sanluri, Segariu, Setzu, Siddi, Tuili, Turri, Ussaramanna, Villamar, Villanovaforru, and Villanovafranca.

38.   Bonifazi, C.; Heins, F. Le dinamiche dei processi di urbanizzazione in Italia e il dualismo Nord-Sud: un'analisi di lungo periodo. Rivista economica del Mezzogiorno 2001, 15, 713–748.

39.   Calza, B.P.; Cortese, C.; Violante, A. Interconnessioni tra sviluppo economico e demografico nel declino urbano: il caso di Genova. Argomenti 2010, 29, 105–131.

40.   Dematteis, G. Montagna e aree interne nelle politiche di coesione territoriale italiane ed europee. Territorio 2013, 66, 7–15.

41.   Author's elaborations on census data from 1861 to 2011. In La popolazione dei comuni sardi dal 1688 al 1991; Angioni, D., Loi, S., Puggioni, G., Eds.; Cuec: Cagliari, Italy, 1997.

42.   Author's elaborations on Data DEMO-ISTAT 2014 and on Urbistat's Data. Available online: http://www.urbistat.it (accessed on 21 November 2014).

43.   Author's elaborations on Urbistat's Data. Available online: http://www.urbistat.it (accessed on 5 March 2015).

44.   For example, the Touristic Consortium of Genna Maria in the municipality of Villanovaforru was the first such association. This story

begins in 1969, but was only legally recognized in 1982 because of the lack of agreement among the different municipalities. The Consortium, that later changed its name to Sa Corona Arrubia, then had members from other municipalities. Today there are about twenty members, and almost the same number of museums, as well as thirty-two archaeological areas, and a park's environmental interest. Today Sa Corona Arrubia is having difficulties, although numerous other Consortia (the Consortium of Two Jars; the Consortium of the Natural Park of Monte Arci; and the Consortium Sa Perda e'Iddocca) have joined it over time. All these Consortia are directed and oriented by , and the interprovincial Local Action Groups (LAG) of the Union of Communities of Marmilla, development agencies; the interprovincial Local Action Groups (LAG) of Marmilla Borgioli C., Deligia M.G., Sa Corona Arrùbia—Consorzio Turistico della Marmilla. Available online: http://sistemimuseali.sns.it/content.php?idSC=103&el=9&c=106&ids= 3&o=sistemiCulturali_dataInizioInterna#ds1 (accessed on 11 May 2015).

45.   Sirena, P. Museo Sa Corona Arrubia Museo Naturalistico del Territorio G. Pusceddu. Available online: http://www.sacoronarrubia.it/ (accessed on 4 February 2015).

46.   In 2005 the World Tourism Organization (WTO), used the "three pillars" to describe sustainable tourism: "sustainability principles refer to the environmental, economic and socio-cultural aspects of tourism development, and a suitable balance must be established between these three dimensions to guarantee its long-term sustainability". Making Tourism More Sustainable: A Guide for Policy Makers; United Nations World Tourism Organization (UNWTO) and the United Nations Environment Programme (UNEP): Madrid, Spain, 2005; pp. 11–12.

47.   Fons, M.V.S.; Fierro, J.A.M.; y Patiño, M.M. Rural tourism: A sustainable alternative. Appl. Energ. 2011, 88, 551–557.

48.   Rural Development Programme (RDP) 2014–2020. Available online: http://www.regione.sardegna.it/speciali/programmasvilupporurale/2014 -2020/psr-2014-2020 (accessed on 8 May 2015).

49.   Regional Development Plan (RDP) 2014–2019. Available online: http://www.regione.sardegna.it/j/v/66?s=1&v=9&c=27&c1=1207&id= 44429 (accessed on 8 May 2015).

50.   Belij, M.; Veljković, J.; Pavlović, S. Role of local community in tourism development: Case study village Zabrega. Glasnik Srpskog geografskog drustva 2014, 94, 1–14.

51. Hall, M.C.; Allan, W. Tourism and Innovation; Routledge: London, UK, 2008.

52. "The structural dimension (network ties, configuration and stability) reveals non-hierarchical and dense ties [...]. Community members' relationships are direct, informal, and long-term. There social ties feed and consolidate a strong sense of belonging to the place where these communities live and work and are the base on which inter-member economic ties and knowledge sharing are developed". Go, F.; Della Lucia, M.; Trunfio, M.; Martini, U. Governing Sustainable Tourism: European Networked Rural Villages. In Rural Cooperation in Europe: In Search of the 'Relational Rurals'; Kasabov, E., Ed.; Palgrave Macmillan: Basingstoke, UK, 2014; pp. 173–174.

53. "From a cognitive dimensional perspective, social capital suffers from the fragmentation inherent in the heterogeneity of local stakeholders and sectorial diversification. Diversity of aims, interests, and competence renders the establishment of a critical mass for group decision-making among the local stakeholders difficult. [...] Information and education could help increase the awareness that networked knowledge sharing facilities inclusive economic institution-building and the emergence of a virtuous cycle of value-adding processes". Go, F.; Della Lucia, M.; Trunfio, M.; Martini, U. Governing Sustainable Tourism: European Networked Rural Villages. In Rural Cooperation in Europe: In Search of the 'Relational Rurals'; Kasabov, E., Ed.; Palgrave Macmillan: Basingstoke, UK, 2014; p. 175.

54. "The relational dimension of social capital focuses on the character of connections, which serve to reinforce not only an organisation's internal logic, but also (external) trustworthy relations". Go, F.; Della Lucia, M.; Trunfio, M.; Martini, U. Governing Sustainable Tourism: European Networked Rural Villages. In Rural Cooperation in Europe: In Search of the 'Relational Rurals'; Kasabov, E., Ed.; Palgrave Macmillan: Basingstoke, UK, 2014; p. 176.

55. Nasser, N. Planning for urban heritage places: Reconciling conservation, tourism, and sustainable development. J. Plan. Lit. 2003, 17, 467–479.

56. Weaver, D.B. Tourism and the elusive paradigm of sustainable development. In A Companion to Tourism; Lew, A.A., Hall, C.M., Williams, A.M., Eds.; Backwell: Oxford, UK, 2004; pp. 510–524.

57. Law, R.; Leung, R.; Buhalis, D. Information technology applications in hospitality and tourism: A review of publications from 2005 to 2007. J. Travel Tour. Market. 2009, 26, 599–623.

58.   Garau, C.; Ilardi, E. The "Non-Places" Meet the "Places:" Virtual Tours on Smartphones for the Enhancement of Cultural Heritage. J. Urban Tech. 2014, 21, 79–91.

59.   Garau, C. From Territory to Smartphone: Smart Fruition of Cultural Heritage for Dynamic Tourism Development. Plan. Pract. Res. 2014, 29, 238–255.

60.   According to Porrello, cultural planning enhances the totality of cultural, environmental, and naturalistic goods by taking strategic actions that are in line with the economic and social sustainability of the local context under study. Porrello, A. L'arte Difficile del Cultural Planning; Grafiche Veneziane: Venice, Italy, 2006; Available online: http://www.iuav.it/Ateneo1/docenti/pianificaz/docenti-st/Antonino-P/materiali-/Cultural_Planning.pdf (accessed on 11 March 2015).

61.   Below are some examples. Available online: http://www.turismo.intoscana.it/site/it/itinerario/Sei-un-turista-rurale/ and http://www.iAGRITURISMO.it. This last link in particular allows users to view the portal of Italian Farms in i-mode, with any type of mobile phone. Here too, the Tuscany region ranks high in the number of farmhouses present.

62.   Mistretta, P. Beni Culturali e sistema territorio. ArcheoArte 2012, 1, 11–19.

63.   Mureddu, D.; Murru, G. Alla scoperta dei monumenti della Marmilla: Archeologia, arte, architettura; CRES: Cagliari, Italy, 2000.

64.   Lilliu, G. La civiltà dei Sardi dal paleolitico all'età dei nuraghi, (Vol. 98); Nuova Eri: Torino, Italy, 1988.

65.   Atzeni, C., Ed.; I manuali delle colline e degli altipiani centro-meridionali; Tipografia del Genio Civile: Roma, Italy, 2009.

66.   Atzeni, C., Sanna, A., Eds.; Architettura in terra cruda; Tipografia del Genio Civile: Roma, Italy, 2009.

67.   Purpura, A.; Ruggieri, G. Il distretto turistico: Caratteristiche e modelli organizzativi. In I distretti turistici: Strumento di sviluppo dei territori. L'esperienza nella regione Sicilia Milano; Cusimano, G., Parroco, A.M., Purpura, A., Eds.; Franco Angeli: Milan, Italy, 2014; pp. 19–41.

68.   McAreavey, R.; McDonagh, J. Sustainable rural tourism: Lessons for rural development. Sociologia ruralis 2011, 51, 175–194.

CHAPTER 14

The Cultural Tourism Management under Context of World Heritage Sites: Stakeholders' Opinions between Luang Prabang Communities, Laos and Muang-kao Communities, Sukhothai, Thailand ⁎

Kunkaew Khlaikaew

Department of Tourism and Hotel, Pibulsongkram Rajabhat University, Muang District, Phitsanulok Province 65000, Thailand

ABSTRACT

The purpose of this qualitative research is to study opinions of stakeholders from Luang Prabang communities, Laos and Muang-kao communities Amphoe, Muang, Sukhothai Province, focusing on the cultural tourism management under context of world heritage sites. The results from the study illustrated that there are 3 types of stakeholders who are involved in cultural tourism promotion and policy formation: 1. Government agency of Luang Prabang and Sukhothai such as, the Heritage House which is responsible for the management of the tourist attractions and town's patrimony in the country under the supervision of UNESCO but for Thailand there is the Tourism Authority of Thailand, Ministry of Sport and Tourism and Fine Art Departments. 2. Entrepreneur includes not only, hospitality, restaurant, and tourism business, but also entertainment business and tourism training business. They are considered to be parts of the support in enhancing cultural tourism to promote the impressive services for visitors. 3. General public understood and agreed with government management. However, the public insisted to participate in preservation of national cultural heritage under the "Live World Heritage Town" concepts and management of UNESCO which Lung Prabang has been protected by these heritage sites since 1995 but Sukhothai has been protected by the heritage site since 1991.

Keywords: Cultural tourism; Luang Prabang city; Sukhothai Province; World heritage sites

1. INTRODUCTION

There are many tourist attractions in ASEAN countries which have been officially registered as World Heritage Sites. Luang Prabang city, Lao People's Democratic Republic (Lao PDR), was the first one of its country which was registered in 1997 as Luang Prabang is dominant from other places. It is distinguished by its nature, society, art, and architecture. Its nature is remarkable because it is surrounded by two rivers passing by the city, and it is situated near by Kong River and surrounded by valleys. Being with nature, the people's life style here is still as simple as the one of the country's life, and it has the features of an agricultural society. Moreover, Luang Prabang has so many exceptional characteristics that it was registered as a World Heritage Site from UNESCO so the number of tourists who have been visiting here, is always increasing. It is the good reason why Luang Prabang can still keep its identity and maintain cultural tourist attractions until now.

For Thailand, the cultural tourism is recognised among Thai people as they focus on the knowledge about culture, custom, life style and many interesting parts of wisdom. The number of people who visited Sukhothai province are 784,555; domestic tourists and 191,956 are those of foreigners. (Statistics Office, The Tourism Authority of Thailand, 2011). Sukhothai Province, Thailand, was the first capital of Siam between the 13th-14th century (around 18-19 of Buddhist era). The city possesses a large number of exquisite edifices that represent the beginning of Thai architecture. The great civilization which had been developed in the Sukhothai kingdom was influenced from the old tradition and culture of the rural area. That had been mixed into Sukhothai's art .The evidence is the Sukhothai Historic Park which is situated at Muang Kao Sub-District, Muang District, Sukhothai province. It is in the east; 12 kilometres from the province, at Lan Ta Pak area and rectangular in shape. There are city walls with the zigzag of three level trenches. In the west, you can see Pratak Mountains, in the east, there is Mae Lam River which flows to Yom River which is about 12 kilometres away.

Nowadays, Sukhothai Province and Kampangpetch Province are deserved as The World Heritage Group. They are ancient cities which have the evidence of the prosperous civilization in the past. All reflect the image of Sukhothai which literally means "Dawn of Happiness". It was the origin of the Thai history which became the important state of South East Asia between 18th -20 th century of Buddhist era. They are protected and developed to be The Historic Parks and registered to be the World Heritage Site in 1991 with the standards of outstanding universal value. There are two criteria for selection that meet the standard, the first one is the site must represent a masterpiece of human creative genius and another one is it meet the third criterion that the site needs to bear a unique or at least exceptional testimony to a cultural tradition or to a civilization which is living or which has disappeared. Lotus bud shape stupa characterises the Sukhothai architectural arts. Moreover, the sculpture of Buddhist disciples walking with their hands clasped together in salutation represents the success of the first era of Thai art.Muang Kao Community (ancient community),Muang district, Sukhothai province, is the nearest community to The Sukhothai Historical Park .Its people have a simple life although they are in the city area. Some areas are the rural community. They still have the agricultural life style. They are outstanding in architecture, society, culture, art and language, all are interesting. However, this community needs the guidelines of development for the cultural tourism management. If the private and public sections, including all people brainstorm together in order to promote Muang Kao Community to have the tourism sources or to set tourism activities at the same old place, the community can become the lively World Heritage, the number of tourist will increase, the community can get more profit. There will be more tourists for the accommodation in Sukhothai. The cultural tourism can alternate the old area of cultural tourism to be the new interesting tourist attraction so that in the future, it can be comparative to Luang Prabang city, Lao People's Democratic Republic.

2. THE PURPOSE OF THE RESEARCH

The purpose of this research is to study the policy on the cultural tourism management under context of World Heritage Sites in the opinions of stakeholders, from Luang Prabang communities, Laos to apply to Muang-Kao Communities, Amphoe Muang, Sukhothai Province.

3. RESEARCH METHODOLOGY

The secondary data was performed by studying data related to the cultural tourism management, the tourism of Lao People's Democratic Republic. The data related to the tourism of Muang Kao Community, Muang district, Sukhothai province by studying the document research such as published media, internet, books, documents and related researches to use these data to make the frame of education.

The primary data was performed by collecting data from the people who are the key informants, both from the people in Lao People's Democratic Republic and from those of Muang Kao Community, Sukhothai province.

3.1 The scope of study The scope of content

3.1.1 Study and analyse the opinions of stakeholders of the cultural tourism management under context of World Heritage sites of Muang Kao Community, Muang district, Sukhothai province.

3.1.2 Study and analyse the opinions of stakeholders of the cultural tourism management under context of world heritage sites of Luang Prabang communities, Lao People's Democratic Republic.

3.1.3 Study and analyze the opinions of stakeholders of the cultural tourism management under the context of world heritage sites of Muang Kao Community, Muang district, Sukhothai province and Luang Prabang communities, Lao People's Democratic Republic.

3.2 The scope of area and samples

Being the world heritage site is the important reason for the researcher to select the studied area. The first one is Luang Prabang community, Lao PDR since it was the first one of Lao PDR that was registered as the world heritage site in 1997 by UNESCO .Another place is Muang Kao Community , Muang district, Sukhothai province which was the first Kingdom of Thailand that was registered as the World Heritage Site in 1991 by UNESCO.

3.3 Populations and sample

The population consists of 5 groups by using the purposive sampling; the cultural tourism authorities. The study focused on the stakeholders of cultural tourism both from Lao People's Democratic Republic and Muang Kao Community, Muang district, Sukhothai province. The 5 groups are as follows:

The first group consists of the executives from 3 cultural tourism travel agencies of Thai and 3 companies from those of Lao PDR

The second group are the 5 repetitive tourists and they used to travel both in Luang Prabang community and Maung Kao, Muang District, Sukhothai province.

The third group are the group of tourism business, which are 5 from Thailand, 4 from Lao PDR.

The fourth group are the scholars of the tourism management; 3 from Lao PDR and 2 from Thailand.

The fifth group are the government officer who have the priority to set the policy; 1 is the mayor of Maung Kao community, 2 leaders of Muang Kao community, 1 leader of Lao PDR and 1 minister of culture of Lao PDR.

3.4 Research Instruments

Semi-Structured Interview was used .The question focuses on the cultural tourism management under context of World Heritage Sites and interview the Key informant by using In-depth Interview with Structured Interview for collecting data in order to know about the concept and the policy of cultural tourism management of both places and analyse for making comparison.

Moreover, the researcher did the Participant Observation by observing the condition of two places and analysed data by the determined topic.

3.5 Analysing and checking information

Data were analysed and checked by using the Data triangulation technique (Jantawanich, Supang,2003). The researchers checked data and classified them by types and by topics and did Qualitative Content Analysis by topics found for finding the relationship and linked data together and extended the result by reporting the result using Descriptive Analysis. Then, the situation of tourism was concluded for analysing and comparing the guidelines of development for the cultural tourism management.

4. CONCLUSION

The research of The Cultural Tourism Management under Context of World Heritage Sites: Stakeholders' Opinion between Luang Prabang Communities, Laos and Muang-kao Communities, Sukhothai ,Thailand, can be Concluded by these following topics.

4.1 The history of Khweang Luang Prabang

In the past, Luang Prabang was the old capital of Laos. The prince Fa Ngoum or Fa Ngum consolidated his kingdom by centralizing the lands of Tai-Lao in the basin of Mae Klong River , Klan River and Ou River with the support of Khmer's king since his wife was the Khmer princess, at that time ,there was the beginning of the acceptance of Buddhism which replaced the belief of spirit. Laos is one of the countries that share the same ancestor as Thailand but Laos has many small tribes. The genuine race has only 50 % who mostly live nearby the Maekhong River but the hill tribes like to live at the mountains. From the beginning Lanchang kingdom was called Chawa since there were many Chawa people who lived here more than other groups. In 1357, the name was changed to Chieng Thong until Khmer's king gave one statue of Buddha which named Pra Blang; Singhol art so King Fagum changed the city name to Luang Prabang and moved the capital of Lanchang to Vientiane ,although Luang Prabang was not the capital but The King still lived in Luang Prabang, then Lanchang kingdom was divided into three kingdoms; LuangPrabang Kingdom, Lan ChangVientien Kingdom and Lanchang Chumpasak Kingdom The king of Lanchang still was on the throne until the end of the dynasty because it was colonised by Siam, Vietnam and French. From the long history, Luang Prabang which was the ancient capital had many temples and beautiful nature. Laos is situated in Suwannaphum or Indochina between 100-108 latitude. In the past, Laos had a large territory. They were mostly mountainous which are 13 important water sources such as Nam Khan, Nam Ngum, Nam Ou ,Nam Sub,

Nampang. The most important and international river is Klong River or its people call Mae Kong. The life style and language of Laos and Thai are similar. Moreover, the cultural festivals between Laos and the north-eastern part of Thailand are not much different. These relate to Buddhism which called Heed Sipsong. Heed means customs. Sibsong means twelve months. Heed Sibsong then means customs that local people practise in each month of year. Various festivals reflect a firm belief in Buddhist religion by people of Isaan region. Heed Sibsong practices, reflect thoughts and ways of life, and is a link that connects and unites people and official symbol of Laos PDR.

4.2 The history of Muang Kao Community

Seven hundred years ago, Muang Kao sub district, Muang District, Sukhothai province, was the centre of the first Kingdom of Thai which filled with the ultimate prosperity from the past in all aspects; politic, religion, art, custom, culture including industry and technology. Thai alphabets were first invented here. There were many parts of local history, art, custom, culture patrimonies; ancients monuments, temples, chedis, all reflecting the prosperity of Sukhothai era. So, at our present time, we call this area Tombol Muang Kao (ancient sub district). It covers an area of about 100 square kilometres. It is in the west of Sukhothai province;12 kilometres from Muang District (city)and the City Hall. The administration was divided into two sections. Subdistrict Administrative Organisation has 11 villages and Muang kao Subdistrict Municipality consists of 12 communities . Muang Kao community is the part of Muang Kao subdistrict, situated in the west of Sukhothai province, the national route 12; Sukhothai-Tak, 12 kilometres from the province. Muang Kao community has shared the same history as Sukhothai around 700 years ago, it is situated at Sukhothai Historic Park and it was the old palace area. People call this area as "Muang Kao" since it was the old capital and local people still keep the old custom and culture until now. Muang Kao sub-district consists of 12 communities which are Trapangtong, PaMamuang ,Ramyai, Ramlek, MaeRompun, Lithai, Sukhothainakorn1, Sukhothainakorn3,Srichum(Jirawat Peerasan, 2011 page 1). The areas of community mostly are basins and there are a few mountains. Clay is the kind of soil found here. There are rivers, Sareetpong Lake and Trapang Dam which keeps the water so that the local people can use the water throughout the year. The tourist attraction is Sukhothai Historic Park which is the great heritage of Thailand. There are many important monuments; more than 14 temples, city pillar, the statue of the King Ram Khamhaeng, brick and clay pottery kiln and citadel etc.

Strengths

1. It's an ancient and important city as it was the capital of Lanchang Kingdom.

2. It possesses long lengthy history.

3.   There were sophisticated arts and architectures.

4.   Its natural tourist's attractions are beautiful and unspoiled.

5.   It was registered as the World Cultural Heritage Site from UNESCO; the world organisation.

6.   The country has its own dominant identity and character presenting Laos ethnic.

7.   There are many world reputable tourism sources.

8.   It has a variety of ethnics so it can be the cultural tourism selling-point.

9.   The people are friendly and have a simple and interesting ways of life which interest tourists

Weaknesses

1.   Its geography is in the kind of land lock; there is no way to transport by the sea.
2.   The climate always changes due to the world climate.
3.   The cost of living is pretty high.
4.   It still has the difficulties to reach to the natural tourism sources, and the facilities for tourists need to be more developed.
5.   The form of tourism still sticks to ancient monuments and objects.

Threats

1.   The government policy of management is always being changed so that the working plan can't be finished.
2.   There is new culture coming with the tourists and that makes Laos culture to get weaker.
3.   The political crisis effects the country administration.

Opportunities

1.   There is international investment for the road and air transportation connecting to ASEAN countries.
2.   The popularity of the cultural tourism interests more tourists to visit Luang Prabang.
3.   Opening the country and attending to ASEAN are the better alternatives to Laos PDR to be developed in all system

4.4 SWOT analysis of Muang Kao community, Sukhothai province

Strengths

1.   It's an ancient and important city as it was the capital of Thai Kingdom.
2.   It possesses long lengthy history.
3.   There were sophisticated arts and architectures.
4.   Its natural tourist attractions are beautiful and unspoiled.
5.   It was registered as the World Cultural Heritage Site from UNESCO; the world organisation.
6.   The country has its own dominant identity and character presenting Thai ethnic.
7.   There are many world reputable tourism sources nearby.
8.   Its people are friendly, and have a simple, interesting life style that interests the tourists.
9.   There are many interesting tourist attractions that can be reached easily.
10.   There are many tourism activities such as porcelain making, Thai desert cooking, the invention from wood.
11.   It has its own Sukhothai language and letters which are the identities of the local area.

Weaknesses

1. Some tourist attractions need to be developed.

2. The location is far from Bangkok; the capital of Thailand.

3. It still has the difficulties to reach to the natural tourism sources, and the facilities for tourists need to be more developed.

Threats

1. The government policy of management is always being changed so that the plan can't be finished.

2. The position of community leaders always rotates therefore the plan can't be done.

3. The climate is always changing due to the world climate.

Opportunities

1. There is international investment for the road and air transportation connecting to ASEAN countries. Opening the country and attending to ASEAN are the better alternatives to be developed in all system.

2. The popularity of the cultural tourism interests more tourists to visit Sukhothai.

3. Sukhothai Province was developed to be the model of sustainable tourism of Thailand.

4.5 The comparison of the cultural tourism management under context of World Heritage Sites between Luang Prabang and Muang Kao community.

The tourism management as the World heritage sites of Luang Prabang ,Laos PDR and Muang Kao community, Sukhothai Province which are managed under the supervision of UNESCO. They are important as the representatives of the culture of communities which represent vividly their own identity and character. Being the source of world heritage sites, they are the important tourist attractions of the country .Being registered as the world heritage, that encourages both foreigners and Thai tourists to flow to visit the historic area and other nearby areas. However ,if the original method of the cultural tourism management from the world heritage source can be developed to be "the lively world heritage". That will encourage Muang Kao community to have the conservation along with the successive development. In conclusion, this research can be divided into 7 topics; tourism resources, accommodation, souvenirs, culture and the ways of life, environments, marketing and information, and the tourism policy under context of world heritage sites focusing on cultural value, social value and aesthetic value. For the issue of the tourism resources in the opinion of stakeholders there were people who involved in cultural tourism promotion and policy formation; government agency, entrepreneurs, general public and the tourists.

Table 1. Tourism resources

The method and the ways to develop of Laung Prabang	The present condition and the guide lines to the development for Muang Kao community
• There are sufficient, correct and impressive of the historic resource information from the local tour guides • The tourist attractions were developed and renovated to the real condition. • The facilities were developed to be sufficient and to have the good quality for long lasting use. • There are campaigns for the conservation of Tourism World Heritage Site • The regulation and the rule of World Heritage Sites are seriously followed. The model of tourism must encourage the tourists. • The program of tourism must respond the need of the tourists. • The program must be created to be more various but it must be under the supervision of the World Heritage Organisation	• There are various ways of giving information; tour guides, audio guide. • The tourist attractions were well developed but there is limited budget. • The facilities are sufficient but only for some point of tourist attractions that are popular. • The Thai tourists need to have more civil common sense, knowledge, and curiosity to historic tourism sources • The people who live nearby the Tourism World Heritage Sites are redundant instead of being part of the valuable tourism resources. • They are many various models of tourism but they are not popular. • It still lacks of the safety in traveling to reach to the remote and dangerous tourist attractions. • The budget is still needed for the development of local resident and need to have the knowledge about the value of the World Heritage Sites • The local organisations are mostly concentrated only to their own communities. • The levels of the participation and integration of working for the World Heritage sites development in all parts are still needed

Table 2. Accommodation

The method and the ways to develop of Laung Prabang	The present condition and the guide lines to the development for Muang Kao community
• The construction of the accommodation must follow the regulation and the rules of the Office of The World Heritage sites. • The accommodation entrepreneurs must follow the rule given by the government under the condition of the Office of The World Heritage sites. • Any kind of construction of the private house must follow the regulation and rules given by the government under the condition of the Office of The World Heritage sites. • Housing condition represents the local identity of Laos. • The design for tourist accommodation is various. It responses tourist need but it still has Laos identity	• There are too much investment for the construction of the accommodation • The styles of accommodation are various since the popularity of modern construction style; therefore, the local identity and character are neither unique and nor harmonised. • There are many styles of housing, traditional and modern. The situation of accommodation is very comparative. It focuses more on the price than its quality.

Table 3. Souvenirs

The method and the ways to develop of Laung Prabang	The present condition and the guide lines to the development for Muang Kao community
• The local products represents its character, the flower pattern textile(one used in the court) • Community products reflect Laos identity. Ban Chang Hai, whisky production • The products represents are not various, there is only the product which had the cultural value • There should be promoted by national organisation so that the products will be qualified	• The local products are porcelain, thai deserts, teakwood cupboard, wood products and local textile • There are not many local products; many of them are from other regions that are sold with local products. • The packages are continuously developed. The symbol was given to promote successively. • The products of Sukhothai province are famous and nationally well accepted such as Sukhothai gold, mineral mud textile(Banna Tonjun) Sukhothai textile.

Table 4. Society, culture, custom and ways of life

The method and the ways to develop of Laung Prabang	The present condition and the guide lines to the development for Muang Kao community
• The ways of life are simple and they are not changing much. • The city zone was clearly arranged such as the old zone and the new zone. • Sticky rice almsgiving morning activity is the advertisement of cultural tourism activity. • The eating life's style still keeps Laos identity : Pho Laos noodle, Kao-peak (Laos noodle), Laos salad. Yet, the creative cuisine is still needed for the restaurant entrepreneurs such as Laos buffet. • The new commercial is reaching the customers ,but here ,they just keep waiting for selling their product to the tourists	• Western culture affects the way of life which became the city's life. • The city's life effects its people, but something can influence the social relationship such as religious ceremony, new-ordained monk ceremony, funeral ceremony, merit ceremony etc. • Having more investments from the capitalist coming from other provinces, the local people are changing the life style; they do shopping in the superstore instead of the market. • Having influenced by the city's life style, the society system is changing. Many local people move to the city.

Table 5. Environment

The method and the ways to develop of Laung Prabang	The present condition and the guide lines to the development for Muang Kao community
<ul><li>Water pollution is emerging, there is more rubbish problems in the city, but the government section has the system to handle the problem.</li><li>The weather is getting hotter due to the global climate change.</li><li>The development may effect to the loss of the forest; trees cutting for huge dam construction.</li><li>Luang Prabang's geography of tourism is still traditional since the whole city is not allowed to the foreign capitalist to do the business such as groceries, shops, restaurant. Laos people must run their own business seriously under the condition of the World Heritage site office. This leads to the ability of saving the original identity which are mixed and harmonised. Therefore, we can conclude that it is the first city that can be the model of the world heritage city in ASEAN</li></ul>	<ul><li>There is less and less water in Rum-pun River.</li><li>The rubbish in Muang-kao is managed by its municipality which is the government organization.</li><li>The weather is getting hotter due to global climate change.</li><li>The environmental management, such as trees in The World Site Heritage, is well managed by Fine Arts Department so the tourism sources are better organized and more beautiful than that of Laos.</li><li>Western culture flows to Muang Kao; there are more convenient stores which are situated among The World Culture Sites and there is no zoning management.</li></ul>

Table 6. Marketing

The method and the ways to develop of Laung Prabang	The present condition and the guide lines to the development for Muang Kao community
<ul><li>The public relations should be done in many ways such as tourism information services by the office of Tourism of Luang Prabang, the given information by the tour guide and the support for tourism by the government.</li><li>Internet trend , globalization and the online link of information are used.</li><li>Word of mouth cab be also the another way of the public relations.</li><li>Television, radio, printing media still play the important role for promote Luang Prabang.</li><li>The campaigns of tourism from the government still encourage tourist to visit Luang Prabang continueously</li></ul>	<ul><li>There are public relations by Tourism Authority of Thailand. Printing media still play an important role for the tourism campaigned.</li><li>Word of mouth can be also the another way of the public relations.</li><li>Internet trend, globalization and the online link of information are used.</li><li>The various concepts of marketing can encourage efficiently both from the domestic markets and those of international.</li></ul>

Table 7. Tourism policy

The method and the ways to develop of Laung Prabang	The present condition and the guide lines to the development for Muang Kao community
<ul><li>It focuses on conserving the area of Luang Prabang systematically, and promotes the trend of lively World Heritage site vividly.</li><li>They tries to build value for the city's life in Luang Prabang, under the concept "conserving and doing for value".</li><li>There should be the support about conservative area successively by having the regulations and law about housing or accommodation under the condition of the World Heritage site and it should be finest harmonised with Laos architecture</li><li>The understanding of local residents in Laung Prabang about the world heritage site</li></ul>	<ul><li>The tourism to Sukhothai is promoted by government section, and the Office of Tourism Authority of Thailand.</li><li>The cultural tourism is promoted.</li><li>The image of Sukhothai translates as "the dawn of happiness" should be promoted.</li><li>Sri Satchanalai and Kamphaeng Phet should be promoted to attract the tourists who visit Muang Kao because Si Satchanalai and Kamphaeng Phet and Muang Kao were registered to be the World Heritage Site simultaneously.</li><li>They should promote the policy about being the closed city so that the tourism of Sukhothai will not be the entertainment place for tourist place for tourist for conserving its own culture.</li><li>There is no policy about tourism management that can link</li></ul>

was created so that they can understand, adjust and accept. • The government section reflects. Being Laos "concept, promote "being proud to be Laos by using the trend of society, culture, the ways of life, and tourism. • Laos People's Revolutionary Party has the policy to foster Lao People's Revolutionary Party has the policy to foster the ways of life of its people.	to other tourist attractions. • There are more support about tourism activity such as the opening of Prasan Market (The ancient local market)

5. DISCUSSIONS

Traveling in the World Heritage Sources is considered as the cultural tourism; the trip to appreciate things reflecting the culture such as palaces, royal courts, temples, ancient monuments, customs, ways of life, arts and all things representing the prosperity which was developed to suit to environments, the life style of people in each era. The visitor will acknowledge history, belief, points of view, faith, and social value of the people in the past which can be passed to this generation by these things. The world heritage sites represents not only the value of the society and the culture in the past, but they also reflect the cultural sites with the power of development, however ,sometimes The culture has no power because the society receives the meaning of culture in the form of powerless. The culture will have the power of development when the understanding of culture occurs. According to Prawet Wasri(1995),the understanding is the great power. Moreover, cultural tourism in the World heritage site should not destroy local art and culture, the tourists should not consider culture as a product, they should support the local culture and do not destroy the local natural environments.

In conclusion, Thailand has many cultural tourism sources which were created by humans for the amusement and the knowledge from the studying about belief, the understanding of the society and culture, the new experience, the civil common sense to conserve the environment and culture. The researcher can conclude that the participation of the government section, entrepreneurs and individuals have to realise one role and duty. If all stakeholders follow the regulations and the rules, the world Heritage Sites can be more jubilant. All should use the creative ways, make the city plan ,create new tourism activities, construct, renovate, change and improve The world Heritage Sites seriously by opening the chance to the residential people to participate in all process; planning, performing and evaluating .In addition, they should share their opinion and manage cultural tourism by using the sources wisely in order to make the balance of the tourism development and the conservation .It is very necessary that the cultural tourism must be done usefully for today and or the future in order to have the sustainable cultural tourism.

6. SUGGESTIONS

Luang Prabang should focus on the creativity of the activities that encourage the tourists to participate for promoting the tourism in Luang Prabang city to be livelier .The bicycles should be campaigned to use in order to reduce the pollution. In addition, travelling by cycling is popular among tourists. The government section should provide bicle lanes and traffic signs, the rubbish problems must be managed urgently, and the government sections should mprove the local food quality for meeting the healthy standard. For Muang Kao Sukhothai, The satisfaction of tour sts must be evaluated. The participation of all stakeholder to the management of the culture tourism must be focuse so that Muang Kao Sukhothai can be the jubilant World Heritage as Luang Prabang .The private sectors should provide the activities that harmonise with the ways of the local residents such as: porcelain, wood carving, learning to cok n odle dishes and the government must have the policy for the construction of accommodations which harmonise to the identity of Muang Kao,Sukhothai for tourists or the tourism entrepreneurs.

REFERENCES

1. Anurak Panyanuwat. (1999) Education and Community Development. Chiang Mai University.
2. Burkart & Medlink. (1981) Tourism : Past, Present and Future. 2 nd. Ed Oxford: Butterworth-Hienemann.
3. Chayan V. (1993) Principle and Theory of Community Development. Bangkok
4. Chusit C. (2001) Tourism Industry, Faculty of Humanities and Social Science. Chiang Mai Rajabhat University.
5. Jiraporn S. (1994) The social history of Phitsanulok 1932-1960 Faculty of Social Science,Naraseun University Jirawat Phirasant , Guide book of Identity of the culture of the communities around the Sukhothai Historical Park. Institute of Mekong-Salween Civilization Studies Naresuan University. Kanokthip P. (1998) Educational Research. Faculty of Education, Chiang Mai University.
6. Kanueng W. (2000) Tourism Relations between Thai-Laos : A Case Study of Thai-Laos Friendship Bridge. Ramkhamhaeng University.
7. Kasem C.. (1987) Environmental science. Bangkok: Auksorn Siam Publishing.
8. Kraisorn W. (2002) Education of personal potential building in the participated project Lao- Europe in three villages of Luang Prabang. Lao People's Democratic Republic. Management of country development. Khon Kaen University.

9. Kunkaew K. (2005) Conservation of floating houses along the Nan River and community participation for tourism. Department of Tourism Industry t. Master of Arts, Chiang Mai University.

10. Mathieson, A., and G. Wall. (1982) Tourism: Economic, Physical and Social Impacts. London: Longman.

11. Office of Tourism and Sports,Ministry of Tourism and Sports (2004) Phitsanuloke Guidebook Phitsanuloke.

12. Ploysri P. (2001) Tourism Planning and Development. Faculty of Humanities, Chiang Mai University.

13. Project Planning section, The Tourism Authority of Thailand. (1997) Ecotourism, Bangkok, The Tourism Authority of Thailand.

14. Sanitwong na Ayutthaya K. (1985) The Conservation and Development in Thailand .The document of Conservation of Nature in Thailand in the case of the development of society and economics. Bangkok: Chutima Publishing.

15. Statistic Office of The Tourism Authority of Thailand. (2011) Tourist Statistic in Sukhothai. Bangkok.

16. Sudaporn W. (1997) Tourism and Heritage Management (International Conference on Tourism and Heritage Management). TAT Tourism Journal, April-June 1997 page 31-33.

17. Supang C. (2003) Qualitative research ,Edition 11.Bangkok: Chulalongkorn University.

18. Thailand Institute of Scientific and Technological Research. (1997) The Operation for setting the policies of eco-tourismSupang Chantavanich. Qualitative research ,Edition 13. Bangkok. Chulalongkorn University.

19. The Tourism Authority of Thailand. (1997) The policy and The Guides to Development of Conservative Tourism of the year 1995-1996. Bangkok References.

20. The Tourism Authority of Thailand and Thailand Institute of Scientific and Technological Research. (1997). The Operation for setting the policies of eco-tourism. Bangkok: The Tourism Authority of Thailand.

CHAPTER 15

Cultural Tourism Potential, as Part of Rural Tourism Development in the North-East of Romania

Florentina Daniela Matei (Titilina)

The Bucharest University of Economic Studies, Patriotilor Street, No. Bucharest, Postcode: 032282, Romania

ABSTRACT

North-East is the largest region of Romania, in terms of number of inhabitants and the area held, but last in relation to the development stage. Although the North-East economy is currently supported by the tourism sector, there is still growth potential, especially referring to business involvement. A qualitative rural tourism requires a process of modernization, development and innovation in the sector by creating competitive travel services in the main areas of interest: Iaşi, Suceava and Neamt. These important North-East cities are loaded with cultural heritage through historical sites, cultural buildings (museums, memorial houses) and religious evidence (churches and ancient monasteries). In order to properly analyze the productivity of the tourism sector, we make reference to the following indicators: accommodation capacity in operation, the number of tourists, the number of overnight stays, average length of stay.

Keywords: rural tourism; cultural tourism; sustainable development; tourism potential, cultural heritage.

1. INTRODUCTION

Favorable conditions , beautiful landscape , air and water purity make from the upland counties of Bacau , Neamt and Suceava , a priceless cultural and religious heritage, therefore North East has a relatively high tourism potential, which can be compared with other popular tourist destinations in the country and abroad.

Tourism activity in this region may be structured as follows (Mihalache, Croitoru, 2006):

Cultural tourism Agritourism Ecclesiastical tourism

Scientific tourism x Therapeutic spa tourism

Agreement tourism

Transit tourism

Cultural tourism

In this paper we want to study the interdependence of the first two forms of tourism mentioned , manifested in the North East region , namely cultural and rural tourism (agritourism) . This will be linked to tourism indicators (number of reception , accommodation capacity and the number of arrivals) and with cultural tourism activity indicators (number and structure of cultural heritage objectives) .

2. METHODOLOGY

This work was done by analyzing the characteristics of rural and cultural tourism and their impact on the evolution of the number of tourists in the North East region.

In order to accomplish this analysis we matched the tourism statistical indicators, the accommodation structure type (farmhouses) and some indicators developed by the Centre for Studies and Research in the Field of Culture detailed as fallows.

We used next statistical indicators (Minciu R.,2000), for the rural tourism sector analysis:

1. The number of existing Farmhouses in North East.

2. Accommodation capacity in operation in farmhouses in North East.

3. The number of arrivals in farmhouses in North East.

4. The number of overnight stays in farmhouses in North East.

Indices that characterize the cultural sector for the regional level are:

1 The optimum use of heritage objectives, by tourism- goals this indicator reveals if cultural tourism potential of each region is used in an optimal way.

2 Cultural tourism index- measuring cultural tourism activities , including road situation in the region.

3 Cultural index- institutional capacity and operators - which measures the level of cultural infrastructure development production and distribution of cultural goods.

Correlating the two categories of indicators will determine the influence of cultural heritage on the rural tourism development level in the North East of the country.

3. NORT-EAST REGION - TOURISM COORDINATED

Region covering the North East of the country and, according to tradition, is part of the old historical region of Moldavia (North East Regional Development Plan 2007-2013). With a total area of 36,850 square kilometers and a population of 3,726,642 inhabitants, the North East is the largest of the eight development regions of Romania. Geographically, the region bordering Ukraine to the north, to the south by the counties of Galati and Vrancea (South East), to the east by the Republic of Moldova and the West Maramures and Bistrita- Nasaud (North West) and Mures counties, Harghita and Covasna (Central Region). With a rich historical, cultural and spiritual region harmoniously combines the traditional with the modern and the past with the present, its potential can be used to develop infrastructure in rural areas , tourism and human resources.

Rural tourism and agrotourism, especially, have great development potential as the region's rural areas have, besides a picturesque landscape, unpolluted and multiple choice recreation and a valuable cultural and historical potential.

Cultural tourism activity , knowledge and information can be structured as follows (Study on the implementation of the Regional Operational Programme in the North- East):

1. Museum tourism supported by a significant number of museums (art, history, ethnography and folklore, technical etc.) , memorial houses that belonged to men of culture , art , science ; fortress , royal courts etc. mention:

Rosetti - Tescanu House (Bacau County) -built in 1898 by Tescanu family - where the great musician George Enescu lived (in this settlement was established opera Oedipus).

Mihai Eminescu Memorial House from Ipotesti- since 1950 has turned into a museum of furniture and other antiques that belonged to the family.

George Enescu Memorial Museum from Dorohoi that exposed personal objects of the artist.

The Palace of Culture built in 1906-1925, is situated on the ruins of medieval royal court. Palace hosts four major museums: the History Museum of Moldova, Moldova Ethnographic Museum, Museum of Art and Science and Technic Museum.

Roznovanu Palace built in Vienna neoclassical style where Iasi City Hall is today.

Vasile Alecsandri National Theatre-built in the period 1894-1896, baroque style decorated, with one of the finest concert halls in the country with a capacity of 1000 seats.

University of Iasi is a former palace that served as the royal court and where worked first Pinoteca in country.

University of Copou is a famous monumental marble hall, known as the " Hall of Lost Steps", decorated with fresco paintings by Sabin Balasa.

Central University Library in Iasi - a building with Doric columns and dome dominating foot Copou.

Royal Inn is one of the oldest civil buildings and houses Suceava Department of Ethnography and Folk Art Museum Suceava County.

City of Suceava built during the reign of Peter I Musat is nearby Bucovina Village Museum, arranged outdoors and containing a valuable collection of traditional construction since XVII-XX.

Ion Creanga Memorial House – Humulesti village (Neamt) contains a number of personal items of great storytelling .

Neamt Fortress , built between 1391-1674 Peter Musatinul withstood sieges over time Austro -Hungarian and Polish armies.

Porumbescu complex museum from Stupca (Suceava) consists of three objectives: the tomb of the great musician, Memorial House and Porumbescu Museum.

Emil Racovita Memorial Museum in Suranesti (Vaslui) is formed from the house where the great explorer was born and it contains documents from the years of school and correspondence with different personalities.

Basil Pārvan Museum of Barlad, found in the former Prefect, raised by Italian neoclassical architects in 1899. The museum contains three sections: art, natural science and history.

Stefan cel Mare County Museum of Vaslui includes sections of archeology , medieval and modern history, ethnography, fine art , and a humor salon – named "Constantin Tanase " in memory of the founder of Romanian theater magazine .

" Silent " Village Museum - founded in 1986, has departments of archeology, ethnography, folklore and so much less to a museum village , rich collection of fine art, rare objects and 400 autographed books by authors. Some exhibits are unique in the country.

Village Museum in Vetrisoaia - ethnographic and historical exhibits.

Cantemir History Museum where there is a collection of archaeological pieces from the Paleolithic to the medieval period, ethnographic collections, fine art collection: George Tatarescu, Dan Hatmanu, Octavian Angheluta.

2. Ethnographic tourism related popular character events (such as the National Festival" Rose of Moldova" in Strunga Iasi festival" Winter Traditions" of Iasi, artisan fairs (" Cucuteni 5000" - Iasi) ; International Folklore Festival" Ceahlaul" (Neamt) International Folklore Festival" Arcanum " (Suceava) International Festival "Hora of elders" (Vaslui) ;

3. Artistic tourism (festivals, seasons, tournaments, exhibitions, etc.); International Festival of Fine Art Tescani (Bacau), Humor Festival of Vaslui .

North East Region has 4043 monuments of international interest, national and local level, according to the heritage list of Culture Ministry in 2004. The list includes archaeological sites and memorial houses, historical buildings and religious monuments. Besides visits to the purpose of pilgrimage or prayer, most tourists visit sights are preferentially oriented religious monuments. They form the "backbone traditional" tourist attractions in the North East. Of the seven tourist areas in Romania covering cultural sites of world importance monasteries in northern Moldavia were included in the UNESCO World Heritage in 1993 and include Voronet, Humor, Moldova, Probota , New St. John of Suceava , Patrauti Church and Tree Church.

Table 1. Heritage list, by county in North-East Region of Romania

County/Region	Local and national historical monuments	Historical monuments of national interest
Bacau	368	95
Botosani	516	46
Iasi	1641	128
Neamt	552	109
Suceava	517	182
Vaslui	449	80
Regiunea Nord-Est	4043	640

Source: Heritage List of the Culture Ministry (processed data), 2004

From the data presented in the table above can be seen as the most developed counties in terms of number of historical monuments are Iasi, Neamt and Suceava.

In terms of number of museums, remains the same counties as the degree of importance, and the total number of museum currently owns 124 in North East Region.

Besides the 3 counties of high importance (Iasi, Neamt and Suceava) Vaslui also stands out with the big number of museum and public collections visitors, In 2013 it attracted 115881 tourits with 10 museums, meaning an average of 317 tourists per day.

Table 2. Number of museums, by county in North-East Region of Romania

Region and counties	2006	2007	2008	2009	2010	2011	2012	2013
North-East Region	124	129	126	127	124	126	118	124
Bacau	21	22	20	20	20	20	20	25
Botosani	11	11	12	12	11	12	10	10
Iasi	24	27	27	27	26	26	21	20
Neamt	28	28	28	28	28	29	28	29
Suceava	30	31	29	30	30	30	30	30
Vaslui	10	10	10	10	9	9	9	10

Source: Statistical Yearbooks of Romania

Table 3. Visitors number in museums and public collections

Region and counties	2006	2007	2008	2009	2010	2011	2012	2013
North-East Region	1849165	1824971	1541777	1685603	1573824	1332343	1336285	1471030
Bacau	45607	46765	54684	61453	51289	58855	60078	76640
Botosani	85074	80636	68985	73214	107507	47237	47116	39802
Iasi	539930	575319	402925	282737	270361	301531	277347	301171
Neamt	433158	512039	427470	625758	609825	378153	400630	437601
Suceava	632169	536325	499522	556883	457787	457712	495130	499935
Vaslui	113227	73887	88191	85558	77055	88855	55984	115881

Source: Statistical Yearbooks of Romania

Considering the number of visitors existing and considering the fact that the study refers to rural tourism, the data obtained above should be correlated with data on agro indicators (Farmhouses number, accommodation capacity, number of arrivals and overnights).

Table 4. Agri-tourism indicators, by North-East Region of Romania

Agri-tourism indicators	Region	2006	2007	2008	2009	2010	2011	2012	2013
Farmhouses number	North-East	177	196	200	241	229	223	265	283
Accommodation capacity in operation in farmhouses		649285	723247	822018	955865	952161	1047315	1246142	1304666
The number of arrivals in farmhouses		52792	64675	74645	77371	73401	94906	102829	105224
The number of overnight stays in farmhouses		94787	117392	133602	141586	134746	177083	202909	193244

Source: Statistical Yearbooks of Romania (processed data)

Applying formulas, it follows that, in the period 2006-2013, the indicators presented the following changes:

- The number of farmhouses increased by 60 %

- Farmhouses accommodation capacity increased by 109%, which means they were built accommodation units with greater capacity than before 2006.

- The number of arrivals in farmhouses doubled in 2013 comparing to 2006, meaning there was demand for new accommodations.

- The number of overnight stays increased at the same rate as the number of arrivals, which means that the average length of stay in a farmhouse remained at 1.83 nights / tourist.

Assuming that the number of visitors in museums and public collections would have decided to stay in rural locations in the North East in 2006, the accommodation capacity could have met only 35% of the existing demand. In 2009 only 57% of these tourists woul had been satisfied with proper accommodation capacity but in 2013 the percentage would had been up to 90%, meaning that agritourism is seen as an opportunity to pastime and also visit cultural objectives in the North-East region.

If we consider that for the same period same of the visitors would have decided to stay in rural locations, the situation would had been as follows:

- In 2006 only 3 % of visitors stayed overnight in farmhouses

- In 2009 , 5% of the tourists could stay in farmhouses

- In the year 2013 the percentage increased to 7% of those people that could stay overnight in farmhouses.

These low percentages may be justified by the existence of a transit tourism that occurs in the NorthEast, the tourist-visitor comes and goes so the accommodation is not a target for them, they just want to visit not to stay overnight or spend a longer period of time in the area.

4. RESULTS AND CONCLUSION

Based on data obtained above we can estimate indices that characterize the cultural sector at the regional level:

A. Optimum use by tour of heritage. The high value of this indicator reflects the intensive use of heritage for tourism activities. Low value indicates reduced their use of cultural tourism.

The calculation: Reporting tourist activity patriomoniu number of targets in each region. Tourism activity was constructed by calculating a score for each region. They consider the following:

• The structures and functions of tourists accommodation

• Capacity and Activity Book

• Number of visitors

• Number of arrivals location.

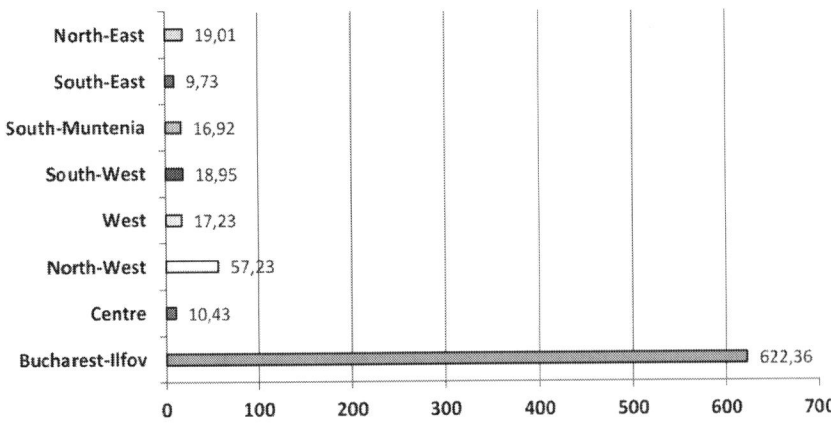

Figure 1. Optimum use by tour of heritage, by regions of Romania

Bucharest-Ilfov Region recorded the highest value because, as the capital of Romania, attracts the largest number of tourists annually. If we don't take into consideration this region, next that uses intensively the heritage elements is North-West (Maramureş) and then North-East.

B. Index for cultural tourism. It is more powerful than optimal use index described above, as stresses and transport situation in the area and the actual number of visitors.

The calculation: Average values:

• Optimal use of heritage through tourism (described above).

• Road.

• The number of visit.

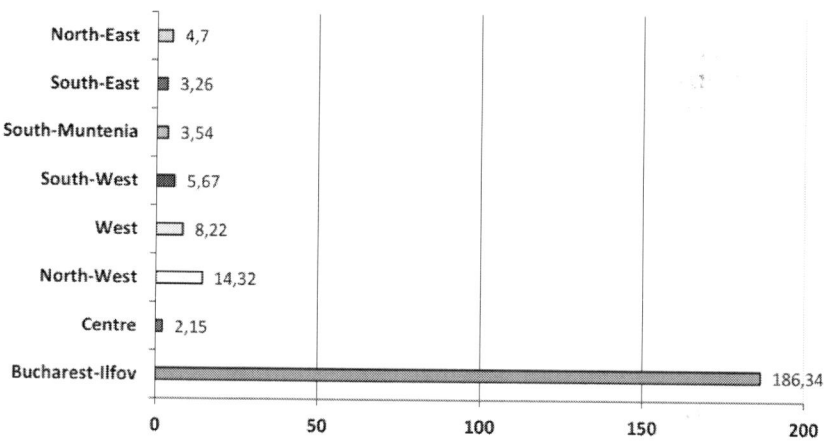

Figure 2. Index for cultural tourism

When you take into account the situation of infrastructure is seen as Northeast Region ranks in the top five since the potential is weak. North West still remains immediately after Bucharest-Ilfov, which means an index of cultural tourism has developed in many ways.

C. Index institutional capacity and cultural operators. High value indicates a developed infrastructure which can support household consumption needs to run programs easier or host cultural events. Low value indicates failure of cultural institutions in the region.

Calculation: This indicator combines a number of cultural institutions in each region / 1.000 inhabitants and combines these results through a medium.

• Heritage objects (number, capacity, hierarchy, types)

• Number of libraries

• Theatres

• Shows Institutions

- Museums
- Cinemas
- UNESCO Monuments

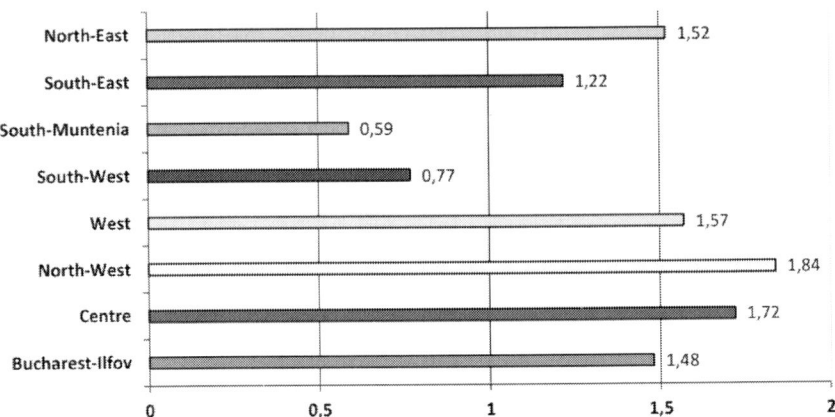

Figure 3. Index institutional capacity and cultural operators

Looking at the chart above you can see that the region can support consumption needs of the population by organizing cultural events is the Northwest, which means it has a large number of cultural institutions based on the number of inhabitants.

In terms of North-East Region, as the 4th in terms of institutional capacity index and cultural operators, the monasteries in northern Moldavia have the most significant role, being included in the UNESCO world heritage.

Tourism in rural areas was, and continues to be increasingly appreciated and requested by people who live and work in more stressful conditions in large urban areas, but not only. Rural tourism is still primarily an opportunity to return to nature, to everything that is pure, unadulterated, a return to origins, always pleasant and refreshing. Tourist activities are alternative income generating activities, which provide opportunities for development in rural areas, due to the unique landscapes, vast areas of semi-natural, innate hospitality of the inhabitants of rural areas. Same time preserving the traditions, culture, gastronomy and rural tourism resources diversity provides a rich and diverse potential for development of this sector.

ACKNOWLEDGEMENT

This work was cofinanced from the European Social Fund through Sectoral Operational Programme Human Resources Development 2013-2020, project number POSDRU 159/1.5/S/134197 "Performance and excellence in doctoral and postdoctoral research in Romanian economics science domain".

REFERENCES

1. Mihalache Fl., Croitoru A. (2006), Romanian rural area: developments and setbacks. Social Change and Entrepreneurship, Expert Publishing, Bucharest, pg. 186
2. Minciu R.,(2000), Tourism economy, Uranus Publishing, Bucharest, pg. 96
3. North East Regional Development Plan 2007-2013- Socio-economic analysis, pg. 91.
4. Study on the implementation of the Regional Operational Programme in the North- East- quantitative and qualitative aspects arising from the economic and financial crisis, pg. 82.

Index

A

approach, 1
ascribed, 1
attention, 1
Authenticity, 181, 182, 185, 186, 190, 192, 193
autocorrelation, 241, 242, 245

B

balance, 5
behaviour, 4
bibliography, 236
broad, 3

C

commercialization, 182, 183, 185, 187, 189, 190
concerning, 6
conclusions, 6
criteria, 5

D

Decentralization, 214
degradation, 11, 13, 14, 15, 16, 17, 19, 64, 120, 197, 232
Demography, 200
demonstrating, 62, 237, 255
determinants, 15
development, 5
dominates, 5

E

elaborates, 197, 255
emphasize, 41, 60, 73, 131, 134, 265
empirical, 23, 24, 27, 183, 184
enormously, 25
Environment, 23
equation, 23

F

factor, 26
frequency, 25
frequent, 27

G

geographical, 39
great, 38
group, 36
Guangdong, 181, 183, 190, 191

H

hard, 42
harmonise, 289
highlight, 39
history, 45

I

identification, 46

inhabitants, 11, 55, 101, 205, 238, 248, 249, 261, 291, 293, 299, 300
inspection, 186
intangible, 46
Intangible, 72, 181
intangibles, 46
Integrations, 214

J

joint, 105
judgments, 109
justify, 77

K

key, 134
knowledge, 158
known, 127

L

limitations, 163

M

Methodology, 229

P

phenomenon, 1, 3, 18, 42, 77, 82, 88, 91, 100, 101, 173, 254, 256

Q

Quality Tourism, 225, 226, 228, 229, 233, 234

questionnaires, 23, 27, 32, 103, 114, 154, 184, 185, 201, 203, 204

R

regression, 190, 244
Rural Tourism, 121, 123, 149, 235, 236, 239, 240, 244, 245, 249, 250, 253, 255, 271, 291

S

social component, 226, 227, 232
Sustainable, 1, 2, 3, 4, 7, 17, 19, 20, 21, 71, 72, 94, 95, 98, 100, 122, 125, 127, 128, 129, 132, 145, 146, 149, 165, 169, 170, 178, 195, 197, 214, 217, 218, 219, 220, 221, 222, 223, 234, 250, 253, 255, 274, 275, 276
Sustainable tourism, 1, 2, 3, 7, 19, 20, 100, 128, 129, 145, 197, 234
Synergies, 211

T

tourism, 1
tourism potential, 150, 163, 257, 291, 292

U

unadulterated, 300
unscrambling, 243